NEW TECHNOLOGY-BASED FIRMS IN THE NEW MILLENNIUM VOLUME VII: THE PRODUCTION AND DISTRIBUTION OF KNOWLEDGE

New Technology-Based Firms in the New Millennium Volume VI (2008)
Aard Groen, Ray Oakey, Peter van der Sijde and Gary Cook

New Technology-Based Firms in the New Millennium Volume V (2006)
Aard Groen, Ray Oakey, Peter van der Sijde and Saleema Kauser

New Technology-Based Firms in the New Millennium Volume IV (2005)
Wim During, Ray Oakey and Saleema Kauser

New Technology-Based Firms in the New Millennium Volume III (2004)
Wim During, Ray Oakey and Saleema Kauser

New Technology-Based Firms in the New Millennium Volume II (2002)
Ray Oakey, Wim During and Saleema Kauser

New Technology-Based Firms in the New Millennium Volume I (2001)
Wim During, Ray Oakey and Saleema Kauser

NEW TECHNOLOGY-BASED FIRMS IN THE NEW MILLENNIUM VOLUME VII: THE PRODUCTION AND DISTRIBUTION OF KNOWLEDGE

EDITED BY

RAY OAKEY

Manchester Business School, Manchester, UK

AARD GROEN

Nikos, University of Twente, Enschede, The Netherlands

GARY COOK

University of Liverpool Management School, Liverpool, UK

PETER VAN DER SIJDE

Free University of Amsterdam, Amsterdam, The Netherlands

United Kingdom • North America • Japan
India • Malaysia • China

Emerald Group Publishing Limited
Howard House, Wagon Lane, Bingley BD16 1WA, UK

First edition 2009

British Library Cataloguing in Publication Data
A catalogue record for this book is available from the British Library

ISBN: 978-1-84855-782-6
ISSN: 1876-0228

Contents

Contributors

Paul Benneworth — Center for Higher Education, Policy Studies (CHEPS), University of Twente, Enschede, The Netherlands

Liesbeth Y. Bout — Capgemini Consulting Services, Utrecht, The Netherlands

Efthymios Constantinides — School of Business, Public Administration and Technology, University of Twente, Enschede, The Netherlands

Gary Cook — University of Liverpool Management School, Liverpool, UK

Sarah Y. Cooper — The University of Edinburgh, Edinburgh, UK

Petra C. de Weerd-Nederhof — School of Business, Public Administration and Technology, University of Twente, Enschede, The Netherlands

Mateja Drnovsek — University of Ljubljana, Ljubljana, Slovenia

Basil G. Englis — Campbell School of Business, Berry College, Mount Berry, GA, USA

Paula Danskin Englis — Campbell School of Business, Berry College, Mount Berry, GA, USA

Aard Groen — Professor of Innovative Entrepreneurship, University of Twente, Enschede, The Netherlands

Gert-Jan Hospers — Department of Human Geography, Radboud University, Nijmegen, The Netherlands

Harmen Jousma — Science and Research Based Business, Leiden University, Leiden, The Netherlands

Jeroen Kraaijenbrink — University of Twente, Enschede, The Netherlands

Kari Laine	Satakunta University of Applied Sciences, Satakunta, Finland
Mirjam Leloux	University of Twente, Enschede, The Netherlands
Jaap H. M. Lombaers	TNO Industrial Technology, Eindhoven, The Netherlands
William A. Lucas	Cambridge-Massachusetts Institute, MIT, Cambridge, MA, USA
Raymond P. Oakey	Manchester Business School, The University of Manchester, Manchester, UK
Igor Prodan	University of Ljubljana, Ljubljana, Slovenia
Elena M. Rodriguez-Falcon	Department of Mechanical Engineering, The University of Sheffield, Sheffield, UK
Victor Scholten	Department of Technology, Strategy and Entrepreneurship, Policy and Management, Delft University of Technology, Delft, The Netherlands
Danny P. Soetanto	Delft University of Technology, Delft, The Netherlands
Michael R. Solomon	Saint Joseph's University, Philadelphia, PA, USA
Peter Timmerman	Studium Generale, University of Twente, Enschede, The Netherlands
Jan Ulijn	Eindhoven University of Technology, Eindhoven, The Netherlands
Laura Valentine	Campbell School of Business, Berry College, Mount Berry, GA, USA
Marina van Geenhuizen	Delft University of Technology, Delft, The Netherlands
Peter van der Sijde	VU University Amsterdam, Amsterdam, The Netherlands
Lorraine Warren	University of Southampton, School of Management, Highfield, Southampton, UK

Chapter 1

Introduction

Peter van der Sijde and Aard Groen

There are many mechanisms by which University knowledge can be exploited by High-Technology Small Firms (HTSFs). This volume contains the best papers presented at the 14th Annual International HTSFs Conference held in the Netherlands at the University of Twente in Enshede, in May 2006, where university knowledge exploitation was its main focus. Although universities play an important role in the upstream generation, transformation, and dissemination of knowledge, university "spin-off" HTSFs have a complimentary role in exploiting the downstream application of these activities. In this exploitation process, the focus of universities is on creating scientific research that is disseminated indirectly (through the educational programmes at undergraduate and graduate levels) and directly to business and society at large. The concept of knowledge exploitation (or technology transfer), therefore, involves the creation of added economic value. However, although knowledge exploitation is often restricted only to those activities that generate economic value, referred to by Etzkowitz (1998) as a "second revolution", knowledge exploitation can also produce social value (e.g., social capital).

In the process of the commercialization of knowledge from upstream to downstream activities, stakeholders are involved in complex relationships, involving academic scientists, business managers, industry scientists, R&D managers, university administrators, technology transfer officers, and entrepreneurs. Indeed, Siegel and van Pottelsberghe de la Potterie (2003) identify numerous barriers to effective commercial knowledge production, including culture clashes, bureaucratic inflexibility, poorly designed reward systems, and ineffective management of university technology transfer offices. None the less, although the development of an effective exploitation policy is often lacking (Bozeman, 2000), exploitation continues to occur!

Knowledge *within* HTSF also needs to be managed and, in the past decade, a knowledge-based perspective has emerged within the strategic management literature (Spender, 1996; Nonaka & Takeuchi, 1995). This perspective builds upon, and extends, the Resource-Based View of Penrose (1959). Also over the years, managerial practice has become more knowledge focused. Often knowledge acquisition is seen as critical for creating and sustaining competitive advantage together with subsequent

New Technology Based Firms in the New Millennium, Volume VII
Edited by R. Oakey, A. Groen, G. Cook and P. van der Sijde

growth, and Alavi and Leidner (2001) provide a comprehensive review of different approaches to knowledge management. However, in general, HTSFs are notorious for staying small, in terms of both number of employees and turnover. In Europe, only a few HTSFs have been able to make the transition from small to large size, although such potential is increased when HTSFs are located in conducive environments such as clusters, incubators, or science parks.

The Exploitation of Knowledge

The presentation of papers in this volume follows the logical and chronological process in which knowledge produced by universities is subsequently exploited. Beginning with a consideration of the business idea recognition process Lucas, Cooper, and Rodriguez-Falcon in Chapter 2 seek to understand the psychological mechanisms that trigger the recognition of science and technology-based opportunities for new ventures. By a number of qualitative interviews at the University of Sheffield, and a survey of science and engineering students at four UK universities, this topic is explored as a major component of opportunity recognition. In Chapter 3, Warren conducts an in-depth examination of the nature and extent of work carried out by a PhD engineer involved in various phases of a high tech university "spin-off." Through a qualitative case study she addresses the important impact that the creation of a university spin-out has on the career development of engineering researchers, which can hinder the spin-out process if not managed effectively.

Jousma and Scholten, in Chapter 4, also report on the role of scientists in the start-up process through a case study of Dutch academic spin-off companies in the Life Sciences. They interviewed scientists who had recent involvement in starting up a "spin-off" firm, based on research they conducted or supervised at the university, and investigated the roles these individuals played in this process. They concluded that a more thorough understanding of the roles that scientists play in relation to the different types of business, the way they are financed, how they grows, and their success in terms of exploitation of the opportunity, can provide useful insights for policy makers.

The Commercialization of Knowledge

This sub-section composed of issues concerned with the commercialization of knowledge is begun by Laine who, in Chapter 5, presents a longitudinal case study of a technology-based enterprise within the context of the Finnish Satakunta University of Applied Sciences. The author analyzes the innovation networks that developed during the first 9 years of firm's existence. The selection of partners appeared to be crucial for the success of this new firm. The substantial knowledge absorptive capacity of the entrepreneur, and his ability to recognize the value of new information, assimilate it, and apply it commercially, enabled him to manage the whole innovation chain. This case study also illustrates that universities should facilitate the embedding of knowledge intensive start-ups within their R&D value network.

Whereas Laine describes a nine-year time span of HTSF development, the authors of the second contribution a focus intensively on one aspect of the commercialization process, namely the valuation of technology as a vehicle for business development. Leloux, van der Sijde and Groen in Chapter 6 first review conventional business valuation methods. The authors then propose a multidimensional approach for business valuation of technology. This approach combines the research value creation perspective (ranging from basic through applied to developmental research), with a business value creation perspective (distinguished by successive business valuation stages). The value of new technologies can be assessed by evaluating changes in strategic capital, economic capital, cultural capital, and social capital. Technology transfer possibilities may yield, for distinctive stakeholders, different outcomes. The authors of the next contribution to this sub-section introduce a tool to improve the commercialization of research findings within a publicly funded Dutch research organization. Bout, Lombaers, Constantinides, and de Weerd-Nederhof in Chapter 7 describe the process of value creation that involves understanding value, creating value, and delivering value. The tool they develop attempts to improve the delivery of value, by organizing a post-project session with key players systematically to discuss the application and commercialization potential of the project under review.

The Management of Knowledge in HTSFs

The selected papers on knowledge management within HTSFs in this section reflect a diversity in knowledge-based perspectives. In Chapter 8, Danskin Englis, Englis, Solomon, and Valentine investigate the process of acquisition, retention, maintenance, and retrieval of knowledge, both within the firm through organizational memory and across the value chain through knowledge management, and compare these practices in small and large firms. Results show that large firms differ significantly from small firms in how they manage knowledge both internally and externally. Larger firms have more developed organizational memory systems. However, small firms are equally as good as their larger counterparts at dispersing organizational memory or sharing information across their organizations. Survey results indicate that smaller firms may not require formal knowledge structures to preserve knowledge. Small size may facilitate informal mechanisms to share internal knowledge. However, in terms of external knowledge management, large firms tend to have more expertise specific to supply chains.

The contribution by Kraaijenbrink in Chapter 9 explores the methods and tools used by high-tech small- and medium-sized firms to identify, acquire, and utilize knowledge from their environment. A survey of 317 small firms showed that they rarely use methods and tools for external knowledge integration. Suggestions are provided as to how usage of methods and tools could be improved by increasing awareness. In the third article on the theme of knowledge management, in Chapter 10, Oakey contributes a paper on the role of HTSF entrepreneurs in using knowledge as a mechanism for gaining advantage over large established competitor firms through the use of new technology to creatively destroy existing technology paradigms. Moreover, this strategic use of knowledge is seen as a means by which HTSFs restore a competitive edge to industry sectors that were becoming dominated by oligopolistic large firms.

Knowledge Support Infrastructures

In Chapter 11, Benneworth et al. present a paper in which two case studies, characterized by regionally engaged universities with regional development strategies, attempt to exploit their regional development potentials. In the North East of England, a partnership of Newcastle City Council, Newcastle University, and the Regional Development Agency (RDA) has jointly purchased a central former brewery site for £30 m (€45 m) on which to develop a new science campus, "Science Central." In Twente, in the east of the Netherlands, the RDA, the University of Twente, its host municipality of Enschede, and a number of other regional bodies have announced support for a 120-hectare science park adjacent to the campus, entitled *Kennispark*. In both cases, national governments have provided moral support, without providing funding for the schemes. In both cases, the schemes are currently under development. In this setting, the authors begin by looking at the novel regional capacities produced by university knowledge commercialization, particularly in terms of the regional networking of HTSFs that have developed around each university. On the basis of interviews, the author develops a new model of community building between high-tech small firms and universities.

In Chapter 12, Soetanto and Van Geenhuizen address a similar topic and focus on socioeconomic networks that support university spin-offs emanating from incubator organizations at the Technical University of Delft as a case study. With regard to support from incubator organizations, they distinguish between conventional (e.g., accommodation, provision of capital) and added value (e.g., business counseling, training) support, and explore whether spin-offs perform better if they receive both types of support compared with only conventional support. The results show that loose networks and weak relationships enhance spin-off growth. It was also confirmed that interaction with partners of diverse backgrounds enhances growth, whereas a positive influence of close geographical proximity with network partners was not important. The results support the idea that receiving a combination of conventional and added value support has a more positive influence on university spin-offs' growth than receiving only conventional support.

Finally, in Chapter 13, Prodan, Drnovsek, and Ulijn present the results of ongoing research into the determinants of technology transfer from academia to new firms. Drawing on previous theoretical and empirical developments in the literature, a conceptual framework for studying technology transfer at the individual's level (i.e., from the entrepreneur's standpoint) is developed. A key element of the conceptual framework, at an individual level, is demonstrated to be the reason why academics seek to become an entrepreneur. The resulting proposed conceptual framework for studying technology transfer from academia to new HTSFs is designed to help researchers, policy makers, and practitioners in designing policy measures to foster this process.

Acknowledgements

We acknowledge the contributions to this introduction from our colleagues, Dr. Ineke Jenniskens and Dr. Ariane von Raesfeld-Meijer.

References

Alavi, M., & Leidner, E. D. (2001). Knowledge management and knowledge management systems: Conceptual foundations and research issues. *MIS Quartery*, *25*(1), 107–136.

Bozeman, B. (2000). Technology transfer and public policy: a review of research and theory. *Research Policy*, *29*, 627–655.

Etzkowitz, H. (1998). The norms of entrepreneurial science: Cognitive effects of the new university–industry linkages. *Research Policy*, *27*, 823–833.

Nonaka, I., & Takeuchi, H. (1995). *The knowledge creating company*. Oxford: Oxford University Press.

Siegel, D. S., & van Pottelsberghe de la Potterie, B. (2003). "Symposium Overview: Economic and Managerial Implications of University Technology Transfer" (Selected Papers on University Technology Transfer from the Applied Econometrics Association Conference on "Innovations and Intellectual Property: Economic and Managerial Perspectives"). *Journal of Technology Transfer*, *28*, 5–8.

Spender, J. C. (1996). Making knowledge the basis of a dynamic theory of the firm. *Strategic Management Journal*, *17*, 45–62.

Chapter 2

On the Recognition of Venturing Opportunities in Science and Technology

William A. Lucas, Sarah Y. Cooper and
Elena M. Rodriguez-Falcon

Entrepreneurial Opportunity Recognition and Directed Attention to Use of Technology

This chapter seeks to understand the psychological mechanisms that support entrepreneurial opportunity recognition. That recognition is treated here from the perspective of Kirzner's (1979) entrepreneurial alertness, a perspective that calls attention to the unconscious processes of discovery. It begins with the proposition that a substantial fraction of opportunities recognised by nascent entrepreneurs are not conscious in the sense that they are not found through a process that is driven by rational search or even by the conscious focusing of one's attention. A cognitive theory of unconscious recognition and discovery is considered that might then explain this phenomenon, based on the proposition that individuals develop a directed attention through interest and experience to quite specific domains.

A summary review of the opportunity recognition literature is offered focusing on the psychology of the recognition process that often precedes conscious search for an idea. The probe into the nature of the opportunity recognition process among a small number of mechanical engineering undergraduates at the University of Sheffield is then offered to provide a qualitative test of the source of ideas, leading to the conclusion that some unconscious process of screening and evaluating ideas must be at work. The next section offers a discussion of the theory of directed attention as a basis for considering what processes might be at play. The Method and Results section then provides the findings from a survey of 494 second, third and fourth year engineering undergraduates at the Universities of Cambridge, Sheffield, Strathclyde

New Technology Based Firms in the New Millennium, Volume VII
Edited by R. Oakey, A. Groen, G. Cook and P. van der Sijde

and York. In the process a measure of Attention to Use of technology will be presented, along with measures of pre-entrepreneurial behaviour and entrepreneurial intent. Concluding comments reflect on the role of university education in the development of unconscious attention.

Opportunity Recognition and Entrepreneurial Alertness

The general view of entrepreneurship that underlies this research is offered by Bygrave and Hofer (1991) who suggest that entrepreneurial research would do well to begin with a definition that the, 'entrepreneurial process involves all the functions, activities, and actions associated with the perceiving of opportunities and the creation of organisations to pursue them'. This definition has proven to be a useful starting point for others (Keh, Foo, & Lim, 2002; Ucbasaran, Westhead, & Wright, 2001), perhaps because it anchors the study of entrepreneurship on a process concept. It also emphasises the role of perception, drawing attention to the cognitive mechanisms that the entrepreneur is said to employ, which may vary from stage to stage in that process. The concern here is the very first step, the recognition of opportunity, and the cognitive processes of entrepreneurial alertness that precede it. The premise is that opportunity recognition includes an unconscious process, in that it is not consciously directed and the individual is generally unaware of their state of alertness. The individual's attention is directed by current interests and intent, and shaped by past personal experience and known information. In this research, the concern is identifying a cognitive process of directed search for new applications of technology for undergraduates studying science and engineering.

Entrepreneurial alertness as advanced by Kirzner (1979) would seem to be an instance of some general form of directed attention. He suggests that the entrepreneur plays a key role in reestablishing market equilibrium by the discovery and grasping of opportunities that others do not see without actually searching for them. 'Entrepreneurial alertness consists, after all, in the ability to notice without search opportunities that have been hitherto overlooked' (Kirzner, 1979, p. 148). It would seem that these discoveries are made through some process of unconscious recognition by agents who are 'spontaneously on the lookout' for unnoticed features in the environment: 'Without knowing what to look for, without deploying any deliberate search technique, the entrepreneur is at all times scanning the horizon, as it were, ready to make discoveries' (Kirzner, 1997, p. 72). Indeed, conscious search is quite difficult if one does not start by knowing what one is looking for, as when Minniti and Bygrave (1999, p. 41) suggest that 'entrepreneurial alertness leads to something previously unimagined'.

Contemporary literature frequently cites Kirzner as evidence stressing the importance of the opportunity recognition (Ardichvili, Cardozo, & Ray, 2003; Kaish & Gilad, 1991; Kirzner, 1979; Shane, 2000). There are, however, wide variations in how much writers on opportunity recognition accept the unconscious nature of Kirzner's entrepreneurial alertness. At one extreme, some suggest that entrepreneurial discovery is the result of classic management techniques that are chosen and consciously directed, abandoning entirely Kirzner's view that entrepreneurial alertness is not consciously directed. Others occupy a more mixed position. Alvarez and Busenitz (2001) seem to include both the conscious and the subconscious approaches but believe the distinguishing characteristic of entrepreneurs is their use of heuristics, defined as simplifying strategies. Although they recognise that entrepreneurs make 'significant leaps' (2001, p. 758), those leaps would seem in their view to follow from a conscious application of decision rules that allow them to deal with ambiguous information and complexity. Studies of the source of entrepreneurial ideas provide evidence of both conscious search and spontaneous recognition, with Koller (1988) finding more opportunities found by discovery than by searching. Herron and Sapienza (1992) suggest that although the mechanisms underlying the search process may be open to debate, it 'apparently involves a subconscious evaluation'; once it 'has synthesized an opportunity, however, conscious evaluation will begin to operate' (p. 52).

Whatever the balance, there appears to be widespread acceptance that entrepreneurial alertness has a strong intuitive or subconscious component. Ardichvili et al. (2003, p. 115) suggest that although conscious search is often important, on balance one should recognise the key role of alertness in recognising opportunities hidden in the information that flows past the future entrepreneur, concluding that whatever it is that happens in a process of passive search is a 'more powerful determinant of discovery — accidental or purposive — than level of activeness of search. Therefore we include entrepreneurial alertness rather than search in our model'. Not dissimilar views are found in Baron (2004) and Gaglio (1997).

A Qualitative Exploration

To explore this notion of unconscious discovery of technology applications, one author has conducted a focus group of undergraduate mechanical engineers at the University of Sheffield. Using questions that were items from a scale on attention to use of technology (see later), the students were asked about how often they had realisations or discoveries about technology, a question important to the quantitative survey methodology reported below. Although the answers varied to some degree, the general consensus was that these students felt that they thought about a wide variety of problems, and about once a month they would spontaneously recognise that some technology might offer a solution. Conversely, when discussing a new technology, they said that about monthly or more often they would seize on either an entirely new application, or more likely a further application of an existing technology. Together these results supported the premise that there is an on-going process of some kind that links problems or needs and alternative uses of technology. Critical to the survey methodology, they were then asked to describe some of those realisations to determine how tangible and consequential they were, and what they were doing at the time.

Some examples drawn from a focus group transcript (Rodriguez-Falcon, 2006) are instructive:

> Samuel: When asked about when a concern for a problem had led to a technology of some kind, Samuel described the experience of not wanting to get up out of bed to change the channel on his television set, setting off the idea of using Blue Tooth technology to link his cell phone to his television. When asked what motivated this realisation, he said 'Laziness, I think'.

Samuel subsequently reconfigured his cell phone to control his television. Later he added remote control of his compact disc player. He now invites friends in and proudly shows his invention.

> Beverly: The realisation she reported occurred while walking a substantial distance to the University on a cold morning that had turned warm by the time she arrived. 'I start feeling hot, so [then I] just start wondering if you could develop a material or fabric that sort of adapts to different climate or temperature changes...' When said she continues to think about the idea from time to time, and adds that, 'Well it's in the back of my memory, maybe subconscious; [I] don't really think on it as much, but, yeah, it's something that I'd like to address, maybe sometime'.

> Timothy: His first recollection of a problem followed by recognising how a technology could help was not particularly impressive: when his electricity bill arrived in the post he recognised he had a problem, and he thought about taking advantage of the more efficient light bulbs on the market. This reminds us that all responses to questions about the frequency of linking technology to problems may be about mundane and well-established technologies, a point worth remembering when we see a large number of instances of recognition of applications reported in the larger survey below.

His second reported application idea occurred later in the interview and occurred against a background that he was reading a novel that featured a group actively opposing any use of nanotechnology. While reading that book, he had from time-to-time thought about how he might use nanotechnology. However, the recognition of a problem–technology linkage was in a conversation with his housemates about painting the inside of his house. After looking at colours they might use, he reported asking what if they could a get a 'paint that changes colour on walls...', envisioning the possibility of a new kind of paint that would have complex molecules that might react to different electrical states controlled by the light switch.

Timothy also reported on an application link recognised on a class team that was assigned to an engineering problem to move an egg from point A to point B. His team had worked at brainstorming together to come up with novel ideas — a form of directed search — but did not reach a satisfactory design. Later, a team-mate went skiing, and after watching the chair lift operate, recognised that an egg-mover could operate on a wire. When he took the idea back, the team successfully applied the idea to their project.

> Herman: He was unable to withdraw money from a bank machine with his gloves on. He took his gloves off, and at the time 'I couldn't hold the card properly and I started shivering and again the slot is so thin so it took me at least two minutes to find the slot'. He has since noticed 'old people as they're struggling to put their cards in the cash machine', and he continues to wonder if a barcode scanner or a chip with user details could be installed instead.

Herman also thought of detachable heels to be used by girls who were taller than their dates, as well as expanding heels for the males. When he and a friend were later watching 'a girl walk by and she was struggling to walk' in high heels, his friend suggested folding heels and Herman returned to thinking about his removable approach.

> Lawrence: Lawrence came to Sheffield to study engineering because of an enduring interest in prosthetic hands, and he had recently seen a new technology 'called nano-muscles, and it uses symmetrical currents to contract'. It came to him that the approach could also be used to communicate sign language, so that among other things a robot could communicate with the deaf. When asked what he was doing at the time he commented that he made the connection in the shower.

There is little evidence of conscious search here. Some ideas are mundane, and undoubtedly many have been discovered previously by countless others elsewhere. We know, for example, of an undergraduate at the University of Ulster and a postgraduate student at Heriot-Watt University who also recognised the value of removable heels. Whatever their economic value, however, the reported instances confirm the presence of an on-going and spontaneous process among engineering students where they from time to time link problems or needs to technology. The ideas seem to arise from a recurring personal experience, from a long-standing interest and/or from an assigned task. The triggers were variously recurring personal need, a new problem or standing in the shower, rather than the result of deliberate search. There, thus, seems to be reason to believe that there is some unconscious process that leads to discovery, and it appears that as a result of their specialised technical knowledge, those discoveries are heavily biased towards the use of technology.

Selective Attention

There is a rich literature in cognitive psychology that supports the premise that individuals are attentive to information both consciously, listening and assessing information with focused attention; and subconsciously, without awareness that they are attending to other sources. A much cited article reported that while individuals varied in their abilities, for the most part a given individual can track around five to seven blocks of related stimuli at the same time (Miller, 1956). As an unavoidable consequence of limited attention resources, the human mind has evolved to serve as a strong filter that simply screens out phenomena not of immediate concern. Broadbent (1952) and Cherry (1953) studied how this subconscious filtering worked in their classic dichotic studies. In these experiments, the subject is asked to wear headphones with left and right earpieces that carry different voices talking about different content. They found that, in general, if the subject is asked to listen with, for example, the right ear, he or she can later describe with accuracy what was said to that ear but will remember virtually nothing about what was directed to the other. A strong conclusion in this and other experiments that have followed is that individuals have a substantial capacity to attend wilfully to one source and that focusing is enabled by the mind automatically filtering out information provided by other senses and sources.

Although we do concentrate our attention resources on some channels and seemingly leave others unattended, it remains that somehow we still notice particular information if it appears in those other channels. The classic example of this process was first cited by Cherry (1953) as the cocktail party effect, where in a crowded and noisy room with many channels of information flowing with information you can suddenly notice that your name was mentioned, and automatically orient and direct your attention to the source. Cherry's dichotic experiments also had the additional finding that there is a similar subconscious process that nonetheless continues to filter the unattended ear and recognises when the attention should be redirected. If the listener's name is spoken into the unattended ear, attention is immediately switched to that voice without conscious effort.

The underlying premise is summarised by Moskowitz (2002) who holds that there is a strong literature that, 'people avoid and approach stimuli prior to recognition, without the individual being aware of the motivational influence'. Note that this is not an ability to be commanded. We are selectively alert to information in a way that is not wilfully directed in a process, and we are not even conscious that it is on-going. (For a review, see Cowan, 1997.) Hence the existence of entrepreneurial alertness as a subconscious process is at least plausible.

The literature is less clear on what conditions cause this process to focus on some and not other information. Given the widespread phenomenon of recognising one's name, the literature question is what conditions are needed for selective attention to operate. This subconscious effect has subsequently been shown to include selective attention to a wide range of 'self-relevant' information, but subconscious attention can also be directed by one's conscious goals (Srull & Wyer, 1986).

It might be supposed that individuals who are strongly motivated and practiced in pursuing technical ideas might be expected to have developed an unconscious alertness to cues that would trigger recognition of entrepreneurial opportunities based on their particular interests and experiences. Those professionally involved in or studying government might be alert to linkages and possibilities involving policy change; managers and business majors would recognise and apply novel marketing ideas in new ways, and science and engineering professionals and undergraduates would, by virtue of their interests and hours spent on technical matters, notice technical solutions — probably tied to their work or course of study, or perhaps to the pervasive Internet technologies that so many are using. If this is the case, then it is likely that such alertness is the result of the training of their attention to filter for and notice opportunities involving the linking of technology and problems or needs.

Method and Results

The processes underlying opportunity recognition are addressed in a study that draws on data collected at the Universities of Cambridge, Sheffield, Strathclyde and York in the United Kingdom at the beginning of the academic year. After filtering out undergraduates who were not in engineering, and the engineering students in their first year (who in October had only just started at the university) and second years who had only one year of university experience, there were 494 completed surveys available for analysis. After a discussion of the measurement of technology alertness, the data are used to test the hypothesis that technology alertness is predicted by technology self-efficacy, venturing self-efficacy and entrepreneurial intent.

Measuring Technology Alertness and other Variables

Technology Alertness

In an effort to measure the alertness concept, items were developed to see if individuals were sufficiently self-aware of instances of when they had recognised a technology-problem linkage, whether they would be able to provide a rough recollection of how frequently they had such recognition events, and whether such answers were random guesses or constituted meaningful data for analysis. Item A, finding oneself wondering in class about how something just learned could be used, and item B, remembering that upon learning a new applied concept they got excited about an application, are at the core of the scale used here (Table 1).

Two other questions were expected to be closely related phenomena. The results from the item about how often these engineering students in the larger study saw something in their studies that could be used to address a social need (Item E) suggest

Table 1: Factor Structure of Alertness to Technology.

How Often Do You...	Frequency More Than Monthly	Component Loading
A. Wonder while you are in class or a lab whether something you just learned could be used to improve a product or process	47.0%	0.744
B. Learn a new applied science concept and get excited about an application idea (whether or not the idea was right)	43.9%	0.737
C. Use a tool or device and it occurs to you that the activity involves some principle you have learned	66.1%	0.713
D. As you learn about a principle, you realise on your own that there are special cases when the principle does not hold up	33.3%	0.695
E. Think about some social problem or need that could be addressed by something you are studying	39.6%	0.642
F. Realise while thinking about a problem that there is technology that could be used in a new way to provide a solution	26.0%	0.636
G. While watching a movie or television, you become very aware that something has violated a science or engineering principle	62.0%	0.610

46.8% of variance extracted. Alpha for seven items = 0.807.

this occurs for 39.6% of these engineering students once a month. A total of 26.0% of students noted that more than once a month they realised while thinking about a problem that there was a technology that could be used in a new way to provide a solution (Item F).

These four types of alertness (A, B, F and E) are the questions presented to the mechanical engineering students at Sheffield. As a set, these episodes of alertness occur monthly or a little less often, at the same rates as found in the qualitative interviews. If one draws a line at a frequency of more than monthly, the proportion that reports a higher frequency varies from 26.0% who more often see new uses of technology to 47.0% who more than monthly wonder while in their classes or laboratories about using what they have learned to make product or process improvements. It is the tangible nature of the examples the Sheffield student could provide that adds some credence to the belief that the instances of discovery and linking are real.

The other statements are a diverse set of questions about the recognition of science and engineering principles in daily life and were expected to form a different scale component.

The result suggests that technology awareness is more diffuse, or perhaps more accurately, less differentiated among these students. All items have a component loading of 0.6 or higher on the same component, and a test of their reliability as a scale yields a satisfactory Alpha statistic of 0.807. A result that suggests the need to develop the scale further is the fact that the factor loadings only extract 46.8% of the variance.

Other Variables

Conceptually one would expect alertness to be higher among students who are confident about their abilities and their intention to be entrepreneurs. The self-efficacy measures follow the work reported by Cooper and Lucas (2006, 2007) that present measures of entrepreneurial intention and self-efficacy. That latter work demonstrates that there are two, separable forms of self-efficacy that can be measured with scales designed to elicit confidence in two different domains. One scale measures confidence in venturing, which is to say entrepreneurship in its more general sense, and is based on a series of judgments the individual provides about their confidence in their ability to, among other tasks, write a business plan, estimate costs of a venture, select a marketing concept and recognise an opportunity. The second scale has to do with confidence in one's abilities in the domain of applied technologies, including the tasks of grasping the best uses of a new technology.

To determine whether entrepreneurial intention drives alertness, we use a scale also developed elsewhere. The items include intermediate and eventual intention, with one item concerned with an opportunity in 'the next few years', and the other open ended, 'At least once I will have to take a chance to start my own company'. For the present study, 20.7% of the undergraduates agreed or strongly agreed on a seven point scale that they would take a near-term opportunity, whereas 23.0% agreed or strongly agreed that they would at least once start a company (Table 2). A similar proportion of 19.6% agreed that a high risk/high pay-off venture appeals to them, and 20.5% agreed that they often think about ideas and ways to start

Table 2: Entrepreneurial Intention Scale.

	Percent Agree or Strongly Agree
If I see an opportunity to join a start-up company in the next few years, I'll take it	20.7%
The idea of high risk/high pay-off ventures appeals to me	19.6%
I often think about ideas and ways to start a business	20.5%
At least once I will have to take a chance and start my own company	23.0%

Alpha = 0.80.

a company. It might be noted that this level of agreement suggests a relatively high level of entrepreneurial intention. When combined in a scale, the Alpha coefficient of reliability is found to be 0.80.

Results

Two background factors commonly found to be predictors of entrepreneurial pursuits are gender and having a father that owns a business. Both are found here (Table 3) to be consequential, with men having higher levels of self-efficacy for venturing ($r = 0.128$, $p < 0.001$) and entrepreneurial intent ($r = 0.196$, $p < 0.001$). The relationship between gender and technology applications self-efficacy is even higher ($r = 0.254$, $p < 0.001$). The reported frequency of instances of technology alertness is also higher for men ($r = 0.234$, $p < 0.001$). Father's entrepreneurial background plays less of a role, although consistent with the literature it relates both to venturing self-efficacy and entrepreneurial intention.

Two other checks on the data are reported for university and year of study. Because the largest number of undergraduates in this study are at the University of Strathclyde in Scotland, which has a university system somewhat different from the three English universities, it seems prudent to see if its students are different on these variables. No differences are found, although it is clear that on average the Strathclyde participants in the study are more often in their fourth year ($r = 0.381$, $p < 0.001$). This leads to a further check to see whether the students starting their third years are in some way different from those starting their fourth year. No meaningful relationships are found between year of study and the other variables and university and year are dropped from further analysis.

The strongest relationship in the study is between two types of self-efficacy (0.646, $p < 0.001$). This result is to be expected, with those confident in one domain likely to be confident in others. A relationship this strong does create an interpretation problem that is resolved below by regression analysis that separates the effects of the two types of self-efficacy on technology alertness. Another expected finding is that venturing self-efficacy is related to entrepreneurial intention (0.245, $p < 0.001$), a finding consistent with the literature.

Our central concern is with the effects of self-efficacy and intent on technology alertness, to test the view that alertness follows from domain-specific confidence and intention. The strongest relationship is found here between technology self-efficacy and technology alertness (0.342, $p < 0.001$), followed closely by the relationship between entrepreneurial intention and alertness (0.303, $p < 0.001$). The relationship between venturing self-efficacy and alertness is somewhat lower, although still quite significant statistically at $r = 0.282$, $p < 0.001$.

As a next step, regression analysis is used to separate out the over-lapping effects of gender, father's business experience, the two types of self-efficacy and entrepreneurial intention on alertness to technology application. Because the units of measurement differ substantially from one predictive variable to another, the

Table 3: Relationships between Background Factors, Self-Efficacy, Intention and Alertness to Technology.

	A	B	C	D	E	F	G
A. Men	–						
B. Father owns a business	0.015 (488)	–					
C. University of Strathclyde	−0.014 (492)	−0.088 (488)	–				
D. Current year (3rd or 4th)	−0.021 (492)	−0.070 (488)	0.381*** (492)	–			
E. Technology application self-efficacy	0.254*** (473)	0.071 (469)	−0.098* (473)	−0.021 (473)	–		
F. Venturing self-efficacy	0.128** (432)	0.140** (428)	0.020 (432)	0.037 (432)	0.646*** (426)	–	
G. Entrepreneurial intent	0.196*** (478)	0.210*** (474)	−0.074 (478)	−0.028 (478)	0.245*** (464)	0.417*** (424)	–
H. Alertness to technology application	0.234*** (483)	0.082 (479)	−0.150** (483)	−0.090* (483)	0.342*** (465)	0.282*** (426)	0.303*** (469)

$N = 494$ third and fourth year engineering students, October 2004.
$^{*}p<0.05$, $^{**}p<0.01$, $^{***}p<0.001$.

Table 4: Regression Analysis.

	Standardised Beta	t	
(Constant)		5.236	0.000
Men	0.151	3.313	0.001
Father owns business	0.019	0.437	0.662
Venturing self-efficacy	−0.025	−0.407	0.684
Technology self-efficacy	0.303	5.175	0.000
Entrepreneurial intent	0.249	5.088	0.000

Multiple $r = 0.490$; $r^2 = 24.0\%$; $F = 25.440$; df = 5, 402; $p < 0.001$.

standardised beta coefficients are provided so one can compare effect sizes. Consistent with the view that alertness is domain specific, technology self-efficacy and entrepreneurial intent are the strong predictors of technology alertness (t = 0.303 and 0.249, both with $p < 0.001$) (Table 4.) The effect of gender decreases when the effects of these other factors are separated out, but it is still consequential (t = 0.151, $p < 0.001$), while a father with an entrepreneurial background plays no role at all (beta = 0.019, not significant).

The striking result is that venturing self-confidence would appear to play no role at all in predicting alertness to new uses of technology (t = −0.025, not significant). Remembering the strong relationship between the two types of self-efficacy, it appears that the correlation found between venturing self-efficacy and technology alertness is spurious, an artefact of their shared correlation with self-efficacy for the application of technology.

Discussion and Conclusions

This research suggests that entrepreneurial alertness as it relates to technology is an unconscious process of recognising linkages and solutions. Once recognised, they would appear to be the subject of conscious attention and evaluation. It seems likely that in a vast proportion of instances the idea is dropped, but the qualitative interviews suggest that some become recurring notions that are elaborated or refocused to test them as solutions to the context at hand. Thus, alertness to technology applications is a domain-specific form of entrepreneurial alertness found among engineers. When the opportunity is tangible and within the resources of the individual, like the mobile phone remote control and the wire egg-mover, the individual acts on the discovery, certainly an encouraging outcome.

The strong relationships found in the regression analysis provide some indication of the origins of this alertness. On the basis of domain self-confidence, one could surmise that alertness is strong when the individual is testing and demonstrating that competence to themselves, and when the occasion permits, to others. It seems reasonable to expect analogous alertness among others. For example, those confident in their sales ability and who intend to pursue sales careers would be alert to opportunities for new sales approaches or channels. One can predict that as education and training, work experience, interests and intentions become more differentiated, the focus of alertness will increasing diversify from one individual to the next.

Whether there is a general form of entrepreneurial alertness is not tested here and requires comparative data. Although the students in this study were found to have relatively high entrepreneurial intent, however, the finding here that venturing self-efficacy among engineers does not increase alertness over that which is predicted by technology self-efficacy is very suggestive. It seems self-evident that public policy would like to see large numbers of engineers who 'get excited about an application idea' and 'realise while thinking about a problem that there is technology that could be used in a new way to provide a solution'.

It is likely that it is technology practice and subsequent enhancement of self-efficacy rather than entrepreneurship courses that strengthen this form of alertness. In that context, the most important thing we do not know from this research is whether the push for entrepreneurship among students detracts from the development of technology alertness in their fields of study.

Summary

This chapter seeks to understand the psychological mechanisms that support the recognition of science and technology-based opportunities for new ventures. Opportunity recognition is viewed as a critical skill in venturing activities, but there is doubt about the mechanisms involved. The entrepreneurship literature contains a tradition of seeing opportunity recognition as a less than conscious process, and certainly not the result of carefully crafted search processes. Kirzner (1997) believes that entrepreneurial alertness is a non-conscious process of recognition; Herron and Sapienza (1992, p. 52) feel that the operation of initial discovery of entrepreneurial ideas 'involves a subconscious evaluation', and Ravasi and Turati (2005, p. 138) consider that the entrepreneurial idea starts with an 'initial intuition'.

Two literatures offer contrasting explanations for how a non-conscious process of entrepreneurial alertness operates. The first is more a personality trait, recognising that large numbers individuals engage fairly deeply with almost everything they are told. Such individuals are said to have a 'need for cognition' (Cacioppo & Petty, 1982). Evidence shows that those with a need for cognition have many attributes associated with entrepreneurs and university-trained scientists and engineers, and with psychological correlates, like self-confidence, one associates with entrepreneurship. A second literature offers an alternative, possibly over-lapping, view that individuals have a learned but unconscious 'directed attention' to potential entrepreneurial opportunities. Individuals who are strongly motivated and practiced in pursuing venturing ideas would be expected to have developed an unconscious alertness to cues that would trigger recognition of opportunities based on their particular interests and experiences. Even among individuals with common science and engineering skills and experiences, some will have an entrepreneurial alertness that would trigger recognition of facts and linkages that involve new venture possibilities that others would not perceive.

This chapter reports on on-going research exploring this second approach, attention to use. Qualitative evidence is presented from a focus group discussion of opportunity recognition conducted with mechanical engineering students at the University of Sheffield. A scale for measuring attention to use of technology is then offered, which includes items used as prompts in the focus group discussion. Results from this scale given to 494 science, mathematics and engineering students at four UK universities are reported, showing that a viable scale can be created, and how this correlated with pre-entrepreneurial behaviour.

The conclusion to the chapter addresses the implications of recognising directed attention as a major component of opportunity recognition.

References

Alvarez, S. A., & Busenitz, L. W. (2001). The entrepreneurship of resource-based theory. *Journal of Management*, *27*, 755–775.

Ardichvili, A., Cardozo, R., & Ray, S. (2003). A theory of entrepreneurial opportunity identification and development. *Journal of Business Venturing*, *18*, 105–123.

Baron, R. A. (2004). The cognitive perspective: A valuable tool for answering entrepreneurship's basic 'why' questions. *Journal of Business Venturing*, *19*, 221–239.

Broadbent, D. E. (1952). Failures of attention in selective listening. *Journal of Experimental Psychology*, *44*, 428–433.

Bygrave, W. D., & Hofer, C. W. (1991). Theorizing about entrepreneurship. *Entrepreneurship Theory and Practice*, *16*, 13–22.

Cooper, S. Y., & Lucas, W. A. (2006). Developing self-efficacy for innovation and entrepreneurship: An educational approach. *International Journal of Entrepreneurship Education*, *4*, 141–162.

Cooper, S. Y., & Lucas, W. A. (2007). Developing entrepreneurial self-efficacy and intentions: Lessons from two programmes. Presented at the International Council for Small Business World Conference, Turku, Finland, June.

Cacioppo, J. T., & Petty, R. E. (1982). The need for cognition. *Journal of Personality and Social Psychology*, *42*, 116–131.

Cherry, E. C. (1953). Some experiments on the recognition of speech, with one and with two ears. *The Journal of the Acoustical Society of America*, *25*, 975–979.

Cowan, N. (1997). *Attention and memory*. Oxford: Oxford University Press.

Gaglio, C. M. (1997). Opportunity identification: Review, critique and suggested directions. In: J. A. Katz (Ed.), *Advances in entrepreneurship, firm emergence and growth* (pp. 139–202). Greenwich, CT: JAI Press.

Herron, L., & Sapienza, H. J. (1992). The entrepreneur and the initiation of new venture launch activities. *Entrepreneurship Theory and Practice*, 49–54.

Kaish, S., & Gilad, B. (1991). Characteristics of opportunities search of entrepreneurs versus executives: Sources, interests, general alertness. *Journal of Business Venturing*, *6*, 45–61.

Keh, H. D., Foo, M. D., & Lim, B. C. (2002). Opportunity evaluation under risky conditions: The cognitive processes of entrepreneurs. *Entrepreneurship Theory and Practice* (Winter), 125–148.

Kirzner, I. M. (1979). *Perception, opportunity and profit*. Chicago: University of Chicago Press.

Kirzner, I. M. (1997). Entrepreneurial discovery and competitive market process. *Journal of Economic Literature*, *35*, 60–85.

Koller, R. H., II. (1988). *On the source of entrepreneurial ideas*. Wellesley, MA: Frontiers of Entrepreneurship Research, Babson College.

Miller, G. A. (1956). The magical number seven, plus or minus two: Some limits on our capacity for processing information. *Psychological Review*, *63*, 81–97.

Minniti, M., & Bygrave, W. (1999). The microfoundations of entrepreneurship. *Entrepreneurship Theory and Practice* (Summer), 41–52.

Moskowitz, G. B. (2002). Preconscious effects of temporary goals on attention. *Journal of Experimental Social Psychology*, *38*, 397–404.

Ravasi, D., & Turati, C. (2005). Exploring entrepreneurial learning: A comparative study of technology development projects. *Journal of Business Venturing*, *20*, 137–164.

Rodriguez-Falcon, E. M. (2006). *Entrepreneurial alertness focus group results*. Unpublished research notes.

Shane, S. (2000). Prior knowledge and the discovery of entrepreneurial opportunities. *Organizational Science*, *11*, 448–469.

Srull, T., & Wyer, R. (1986). The role of chronic and temporary goals in social information processing. In: R. Sorrentino & E. Higgins (Eds), *Handbook of motivation and cognition: Foundations of social behavior* (pp. 503–549). Guildford CT: The Guilford Press.

Ucbasaran, D., Westhead, P., & Wright, M. (2001). The focus of entrepreneurial research: Contextual and process issues. *Entrepreneurship Theory and Practice* (Summer), 57–75.

Chapter 3

Entrepreneurial Identity, Career Transformation and the Spin-Out Process

Lorraine Warren

Introduction

The purpose of this chapter is to explore the creation and maintenance of entrepreneurial identity during the establishment of a high-tech university spin-out (USO) company in the United Kingdom. The chapter is based on a case study of a mature PhD student working initially in a research team, and later in a spin-out company, in a UK university; he then founds his own company. The study tracks his understanding and development of different aspects of his professional identity as he works towards shifting career goals in different formal and informal learning settings. The chapter commences with a discussion of the career tensions that might arise during the spin-out process. The next section argues that purposeful construction of entrepreneurial identity may be a significant element in supporting successful career transformation. The third section presents the case in detail. Following a discussion, conclusions are presented. The practical implications of the study are that better understanding of these processes can be used by educators and support staff in classroom settings and in incubators. Theoretically, the chapter adds specifically to the growing literature on entrepreneurial identity, extending it to the realm of science and engineering; the importance of the dynamic between engineering identity and entrepreneurial identity during the transition from engineer to entrepreneur is examined in depth. Most significantly, the case demonstrates how a complex reworking of what it means to be an entrepreneur and what it means to be an engineer takes place that is enormously significant in the crafting of a personal career trajectory.

To set the chapter in context, a brief overview of the case history is now presented. The engineer in question, Jens, was born in Germany in 1970. In 1991, after school,

New Technology Based Firms in the New Millennium, Volume VII
Edited by R. Oakey, A. Groen, G. Cook and P. van der Sijde

and military service, he studied mechanical engineering to first degree level, followed by a PhD in a UK university between 1997 and 2001, on Intermittent Electrical Discontinuities in Tin-Plated Automotive Contacts. In late 2000, as his PhD was coming close to completion, he began work in his university department for a new spin-out company as an R&D engineer and software developer. His duties were the design of mechanical and electrical hardware as well as software for surface metrology applications (optical non-contact measurement of precision surfaces). From September 2004–2005, Jens also undertook a Masters course in entrepreneurship with a strong focus on business venturing. He has now (mid-2007) set up his own company in the same field and in late 2005 has attracted first round funding from an investor and a UK government R&D grant.

Creating a USO: Challenges to Career Development

There is considerable interest at policy level in the United Kingdom in the creation of wealth from the commercialisation of university research, the Higher Education Innovation Fund being just one of a raft of initiatives to support activity on this area. The spinning-out of new technologies through the formation of USO companies is one mechanism by which commercialisation might occur. Stating the obvious, at some point, if a USO is to be formed, someone, somewhere, has to take the critical decision that forming and growing the new USO will dominate their career path for at least the short and medium term; this is an entrepreneurial role, and for those who have not been down this path before, they must, *ergo*, become an entrepreneur.

Of course, the literature on why anyone starts a new venture is vast, covering issues at the macro level, such as policy, governance and the economic environment, as well as at the meso and micro levels of analysis; furthermore, research approaches have spanned the disciplines of economics, psychology and sociology. The full domain is too broad to review here in any meaningful way; however, it is possible to demonstrate that there is still scope for further research that takes the academic entrepreneur as the focus of analysis, to generate new insights into where educators and support staff might be able to effect improvements in performance. Despite much support and ongoing research and evaluation, it is claimed that the United Kingdom has at times not maximised its full potential in the commercialisation area (Bray & Lee, 2000; Nicolaou & Birley, 2003a, 2003b; Vohora, Lockett, & Wright, 2004), and this chapter makes a small contribution to that practical agenda, as well as making a novel contribution to the literature on entrepreneurial identity.

In a study on social networks and USOs, Nicolaou and Birley (2003b) note that the majority of the literature on spin-out formation has concentrated on issues of patenting and licensing; economic implications; the economic implications of technology transfer; the implications of the Bayh-Dole act and, similarly, changes in the role of the British Technology Group; structure and policies and the significance of internal and external mechanisms in promoting technology transfer. Similarly, in a comprehensive review of the spin-out literature, Djokovic and

Souitaris (2008) also note the predominance of such macro and meso level studies on economic outcomes, technology and market factors, university support mechanisms and network effects. At the micro level, they found that studies typically debated the effect of the involvement and role of academic and/or surrogate entrepreneurs on the performance of spin-outs. These authors call, *inter alia*, for a greater focus on the academic entrepreneur as the unit of analysis. Although they do not focus on career progression and the initial decision to venture as specific issues, Lam (2007) reminds us how little is known about the complex world of scientific careers located in knowledge-based networks, drawing on a study of university–industry partnerships.

It is noticeable, as Audretsch and Erdem (2004) point out, that there is little on why *specifically* engineers or scientists, as distinct from other entrepreneurs, choose to become entrepreneurial (or not). However, while we may wish to focus on the micro level to explore this question more deeply, we must remain aware that entrepreneurial activity does not take place in a hermetically sealed individual vacuum. Although early research on entrepreneurial propensity focussed on individual traits such as need for achievement, attitudes to risk taking, locus of control and so on, this approach has been superseded in the entrepreneurship literature by approaches that take account of wider environmental factors. Of course, government policy has implications (Storey, 1994) though research in this area tends to indicate broad trends rather than individual decisions. With regard to technology specifically, Casson (1995) suggests that an economic context where technological advance is readily embraced in itself presents a more intense demand for associated entrepreneurship, and this may therefore act as a powerful attractant for individuals pulled this way through the potential of high earnings. Anderson and Chorev (2006) explore how engineers learn to become entrepreneurs in Israel through an analysis of push-pull factors that span individual aspirations and motivations as well as the cultural and historical context, while noting that the uniqueness of that setting has a distinctive impact on the aspirations and opportunities in their sample. Chell and Allman (2003) have examined the motivations and intentions of technology-oriented entrepreneurs in structured learning settings, indicating the importance of the individual, their cognitive, behavioural and emotional dimensions in interplays with the broader contextual milieu. They suggest that entrepreneurial metamorphosis is complex and difficult to predict for any individual case. Luthje and Franke (2003) also note the multi-dimensional relation between personality, attitude and barriers and support factors on the entrepreneurship-related context for engineering students at MIT, where an entrepreneurial career is valued as a high prestige option. Meyer (2003) too examines the relation of the individual to the wider context but concludes that public support mechanisms and incentive structures may not necessarily promote academic entrepreneurship. Instead, entrepreneurial behaviour patterns may arise where scientists in public sector organisations are not necessarily interested in setting up a high growth company but are instead looking for an alternative mechanism to support a career where research interests are paramount. Overall, the USO domain is a complex milieu. Perceptions and choices, careers present and future are the result of a dynamic relation between the individual and the cultural, social and technological context that is not well understood.

Certainly, the spin-out domain is not as straightforward for aspirant academic entrepreneurs as early studies might have led us to expect (Druilhe & Garnsey, 2004). These authors have noted the tendency to view academic spin-outs as an undifferentiated category, with spin-out creation conceptualised as a linear process where a technology-based idea is generated from research, protected by patents and transferred to a new firm established to commercialise the idea. However, personal motivations, business competencies of scientist-entrepreneurs, the availability of external resources and the university environment are all found to play a significant role in encouraging or preventing entrepreneurial activity in universities. Not surprisingly then, in light of all those variables, the structure, form, objectives and outcomes of USOs vary significantly according to a number of factors including the status of the individuals carrying out the business venturing process and the nature of the knowledge or technology being transferred (Pirnay, Surlemont, & Nlemvo, 2003); this is certainly the case for Jens as we shall see later. Of course, different USO models present different challenges and opportunities to academics. Nicolaou and Birley (2003a) show that the process and outcomes of spin-out are contingent at least in part on the role and degree of involvement of key academic staff. They propose a trichotomous categorisation of USOs into orthodox, hybrid and technology spin-out. An orthodox USO involves both the academic inventors and the technology spinning out from the institution; the hybrid situation refers to the technology spinning out and the academics retaining their university position, but holding a part-time position within the company, such as a directorship; a technology spin-out involves the technology spinning out, but the academic maintaining no substantive connection with the newly established firm (beyond, perhaps, an equity position, or the provision of consultancy advice).

Nicolaou and Birley (2003b, p. 1704) note that inherent in this trichotomy is the requirement that, at the individual level of analysis, every academic inventor must make a critical career choice. On the one hand, the inventor may leave the university to completely focus his or her energy in the firm (academic exodus); she/he becomes an entrepreneur. On the other hand, the inventor may decide to remain in the university and may or may not accept a part-time position in the firm (academic stasis). Of course, the career choices for those in temporary positions in research units, PhD students such as Jens for example, may well be a little less clear cut. In this case, the choice is between developing the USO, obtaining temporary research funding to continue 'pure' research, obtaining work elsewhere in the university or leaving the university: not between a 'secure' academic position and the USO possibility. Whatever the starting position of the individual in question, PhD student or a fully tenured academic, if the choice is to follow the USO route, this presents a set of personal and career challenges and opportunities to the aspirant entrepreneur. Of course, a spin-out is attractive as it can potentially provide high revenues and prestige both to universities and to the founding academics (Bray & Lee, 2000), but the challenges are multifaceted and change over time as Vohora et al. (2004) have argued.

Vohora et al. (2004) identify five phases of growth for high-tech USOs that highlight the challenging dynamic between the individual and the broader context during the trajectory of a new USO: the initial research phase, the opportunity phase, the pre-organisation phase, the re-configuration phase and the sustainable high growth phase. To reach full potential, it is argued that the venture must successfully make the transition between the phases, overcoming what are termed 'critical junctures' as they move from one phase to the next. The critical junctures concern the absence of key resources or capabilities required by the firm, some of which are tangible business necessities, such as finance, others are less tangible, associated with the personal motivation and career intent of the potential academic entrepreneur. The latter appear to be most significant in the juncture between the opportunity recognition phase and the pre-organisation phase and the juncture between the pre-organisation phase and the re-configuration phase. It is these less tangible capabilities associated with the entrepreneur that are the concerns of this chapter, as they include the decision as to what career path to follow, whether to become an entrepreneur or instead to follow another commercialisation route and let someone else take on that role. Using Vohora et al.'s model, it is possible to focus on the individual level of analysis, the academic entrepreneur and their choices, while not losing sight of the broader context in which that behaviour takes place over the life trajectory of a USO.

Turning to the first phase transition, 'entrepreneurial commitment' is the juncture that is most significant in the move from opportunity recognition — the idea that a new technology may have commercial potential — to the pre-organisation phase. It is in the pre-organisation phase that many fundamental uncertainties about the context, including the industry, location, size, market and team issues, are resolved. Strong intention and commitment are necessary to move forward from a vision to an operational business engaged in commercial transactions (Erikson, 2002). Vohora et al. identify possible challenges to entrepreneurial commitment at this critical juncture:

- institutional culture prioritising research,
- the challenges of an alien commercial environment,
- founding a USO is risky, engineers/scientists may be risk averse and
- difficulties in delegating to business specialists.

At this point, the academic may decide on exodus and found the USO themselves, if she/he has a strong attachment to the possibility of becoming an entrepreneur and is willing to take on that career risk in light of the commercial opportunity seemingly at hand and the nature of the institutional culture. Alternatively, a weaker attachment to the USO may take place; exodus may still occur, but the academic may remain in a research role, relying on other team members to carry out business functions (as 'surrogate entrepreneurs').

Turning to the second phase transition, once commitment has been achieved, then there is the issue of credibility, the juncture between the pre-organisation phase and the reconfiguration phase. This is the 'building' phase for the organisation, the choice for growth or existence; a choice for growth is dependent on presenting a credible businesslike front to customers and financiers. To some extent, going for growth is a functional business issue that can readily be addressed through support in the development of feasible business models, plans and effective presentations, but legitimacy and trust issues also impact at the personal level (Lounsbury & Glynn, 2001) in dealing with sceptical customers and investors. Does the aspirant entrepreneur 'look the part', by taking on the discourse of the customer and the market, or is confidence undermined by a continuation of academic attributes and values that frighten investors away?

Vohora et al. (2004) present evidence that it is the pre-organisation phase that represents the steepest learning curve for the academic, particularly if they have little or no commercial experience or knowledge of how the target industry operates. The entrepreneurial learning has to take place in the context of contemplating a significant career shift that carries with it considerable career risk in terms of university cultures that prioritise research output rather than commercial activity as well as business risk. They also argue that it is the re-configuration phase that presents the steepest learning curve for the entrepreneurial team, where the academic has to manage a new team role, whether as an entrepreneurial leader or not. Of course, this latter depends on the goals of the reconfiguration phase.

Mainstream tenured academic staff who choose a high degree of involvement in a USO are inevitably moving away from a linear model of classical academic career development to a less traditional, multi-faceted path where entrepreneurship is more central, at least for a time. This is challenging, but work is not just about financial reward; it is a source of personal identity and self-fulfilment (Baruch, 2004). Politis (2005) too argues that entrepreneurs have diverse career motivations and that self-image is an important element in the conceptualising of decisions and motivations surrounding career choice. While Baruch (2004) notes that the academic career has a certain fluidity of its own, nonetheless, the academic entrepreneur has to carefully balance their USO aspiration — stasis or exodus — against organisational and external perceptions as well as internal goals. Self-image too has an external dimension in terms of the identified need to appear credible and businesslike to potential investors and customers: should this be delegated to other team members, the inventor academic remaining in the 'back room'? If not, what does 'credible and businesslike' mean? What does it mean to 'become an entrepreneur' in this specific context? Will my engineering reputation be compromised?

From Jens's point of view, his career development prior to spin-out activity has been steeped in the culture of academe and engineering. Considering moving to an entrepreneurial career position requires a reworking of internal values and beliefs to meet his own aspirations and institutionalised sets of expectations of the university and the commercial world. In this chapter, I contend that crafting an entrepreneurial identity may be part of effecting a successful career transformation — but what does that mean for a PhD engineer? The next section underpins this argument.

Formulating an Entrepreneurial Identity

In this section, I examine Down and Reveley's (2004, p. 236) 'the social formation of the entrepreneurial self', from the point of view of career transformation. Lounsbury and Glynn (2001, p. 554) note that a key challenge for entrepreneurs is 'to establish a unique identity that is neither ambiguous nor unfamiliar, but legitimate', arguing for the importance of formulating an entrepreneurial identity for self and firm in acquiring legitimacy in the early stages of venturing. This stance is in line with an empowered vision of entrepreneurial identity (Lounsbury & Glynn, 2001; Down & Reveley, 2004; Reveley, Down, & Taylor, 2004; Downing, 2005; Warren & Anderson, 2005; Down, 2006) where entrepreneurs are acknowledged as skilled cultural operators manipulating perceptions of the entrepreneurial self to achieve desired outcomes for their new ventures. While authors such as Ritchie (1991), du Gay (1996), and Cohen and Musson (2000) discuss the extent to which individuals are reflexively inscribed as entrepreneurs, an identity 'on offer' within the discursive medium of the enterprise culture, nonetheless, entrepreneurial identity is still constituted as a relation between the individual, culture and society in some way. Significant in this view is the notion of identity as a process of becoming (Giddens, 1991), where there is the possibility of agency, and individual identity can be negotiated through and within the sense-making systems of the surrounding cultural milieu (Jenkins, 1996). Thus, self-identity can be crafted and re-crafted as an ongoing project of the self (Giddens, 1991). This clearly has resonance with Baruch's (2004) metaphorical conceptualisation of 'career' as a 'life journey': the emergence of the short-term portfolio career over the past two decades has placed the emphasis far more on the individual in terms of maintaining expertise and employability. In the past, the emphasis was more on the linear progression through an organisation, which in a sense 'owned' the career trajectory. Of course, Baruch (2004) also notes that individual career progressions do not take place in isolation; they are shaped by organisational structures, cultures and processes. Thus, an individual crafts and recrafts their career as a significant dimension of self-identity, in line with Giddens' (1991) theories of structure and agency.

Goffman (1959) emphasises the importance of roles in shaping identity. He describes how individuals 'work' their roles in relation to social expectations; here, identities and meanings are fluid, that is, negotiated and sustained in shifting role-based interactions with others. Goffman also argues, however, that while there is fluidity, roles become institutionalised sets of social expectations, with stereotypes emerging as a more fixed form of meaning and stability. One might consider 'academic', 'engineer' and 'entrepreneur' as career 'roles' that have associated professional expectations; those aspiring to such career roles have to craft their identity to social and professional expectations, with the transition to founding a USO being a time of great fluidity in identity terms.

Summarising the previous two sections, it has been established that:

- some academics may form a strong attachment to entrepreneurship, choosing to find a USO themselves; this is a critical career choice (Nicolaou & Birley, 2003a, 2003b);
- this choice results in challenges at the personal level that present a steep learning curve if the firm is to progress (Vohora et al., 2004), including building legitimacy and trust for self and firm (Lounsbury and Glynn, 2001) and

- Legitimate entrepreneurial identity can be crafted as part of an ongoing project of the self (Baruch, 2004; Giddens, 1991; Goffman, 1959).

Jens clearly makes a journey from being a researcher engineer in an academic setting, to 'becoming an entrepreneur', at the helm of his own new technology company. This case study provides an empirical exploration of the issues summarised above. The focus is on Jens' workings of identity during his transition from academic engineer to entrepreneur, specifically the dynamic between the career roles of academic, engineer and entrepreneur.

Becoming an Entrepreneur?

Methodology

The research phase of this project was designed in an inductive and exploratory manner to obtain a rich understanding of how Jens recognised, considered and assessed issues of identity during his career generally, but more specifically, during his multi-faceted engagement with USO activity (Eisenhardt, 1989). I first met Jens when he was a student and I was a tutor on a Masters course emphasising business venturing. I thought he was a fascinating subject for a case study because of his strong engineering background and because he was so actively engaged with USO activity, that is, in the space where the USO evolves from research activities to a commercial organisation (Van de Ven, 1992). Additionally, I had a supervisory relationship with him during the latter part of the course, which enabled space for intensive discussions. This closeness enabled a speedy rapport between myself and Jens, especially as I too have a scientific background and have worked with USOs in a similar field for 5 years. The disadvantage of this closeness is, of course, bias in the data collection and the analysis that impacts on the generalisability of the study. Hence, as well as material gained from one-to-one discussions between Jens and I, the study was augmented by consideration of reflective material prepared by Jens for other members of the academic staff during his Masters course and material from the company website.

The first tranche of data was collected over a 1-year period from September 2004 to 2005, during the time when he was undertaking his Masters course. During that time, he moved from his initial position with an existing USO to establishing a new USO with colleagues from his university department. A 90-minute semi-structured interview was carried out in March 2005, during which extensive notes were taken; other sources used were, with his permission, Jens's assignments for the course (in which self-reflection was a significant component), notes taken from meetings between Jens, his project mentor, another member of the academic staff and myself and comments from emails. Informal chats then took place during the remainder of 2005. A further intensive period of interaction took place in early 2006 prior to my presenting early outcomes of this project at a conference. During the remainder of 2006, we had informal discussions at networking events; finally, a further 90-minute semi-structured interview took place in 2007, after he had moved on to find his own new technology firm. The analysis of this rich set of data employed the method of reflecting on the data as it emerged, to look for patterns and themes concerning issues of identity, which were discussed with Jens and then related to an explanatory framework of the business' development over time to support conclusions. Of course, there is the potential for subjective bias in the interpretative phase here, in Jens's interpretations and my own. Nonetheless, there is a trade-off in the depth and richness of the material obtained.

Analysis

The analysis is divided into two parts. Firstly, Jens's phases of engagement with USO activity are defined in terms of Vohora et al.'s (2004) model; secondly, his perceptions of personal identity in relation to career issues are analysed in accordance with that categorisation. His engagement spans four of the phases identified by Vohora et al. (2004) and concerns two companies, A-tech (the firm he joined towards the end of his PhD) and B-tech (the firm he set up during this study). This transition is set out in Table 1.

In A-1, he is undertaking a PhD in traditional academic mode, but in a research group where a senior professor (his supervisor) is forming a USO, which presents potential possibilities for his own career future. He also undertakes a German 'Economics for Engineers' distance learning course, with a view to gaining business skills for the future (not necessarily the spin-out company) but leaves because it is 'too theoretical'. In A-2, he completes his PhD and joins the company as it forms in 2001, very much in 'engineer/researcher' mode. In A-3, he is fully engaged with

Table 1: Phases of Jen's engagement with company development activity.

#	Phase	Date	Company	Jens's Activities
A-1	Opportunity	Pre 2001	A-tech	Academic researcher/PhD student/associated with activities of nascent USO
A-2	Pre-organisation	2001–2003	A-tech	Academic researcher R&D software, building software portfolio for company A
A-3	Re-configuration	2004–2005	A-tech	Company R&D/ entrepreneurship student
B-1	Pre-organisation	Late 2005	B-tech	Founder of new company
B-2	Re-configuration of new company (B)	Late 2005 to present	B-tech	Building customer relationships for new company

A-Tech, but still as an employee, not a shareholder. The company is successful, but not growing very quickly. He again seeks to improve his understanding of the commercial aspects of the business outside engineering by joining a Masters course in entrepreneurship with a strong practical focus on new ventures. In B-1, he thinks he can achieve more in his own right, so he leaves A-Tech and finds his own company, in a related but different field: he is developing prototype instrumentation to measure radius, surface roughness and surface defects or wear in spherical objects in the field of medical devices, such as artificial joints. Currently (B2), he is building a market base through developing key customer relationships.

During these phases, Jens is working towards different career goals in formal and informal settings. I now analyse his perceptions of self and career during these phases drawing largely on an interview that covered the history of each phase. It is *particularly important to note* that throughout all the phases, Jens sees himself first and foremost as an engineer and that in any team or business development, he would build personal skills and new relationships to bolster any perceived weaknesses, but most firmly:

> *I would never compromise my reputation as an engineer*

This theme recurred constantly. When asked what he considered to be the prestige factors for an engineer, the following were suggested:

- good engineers can not only make something work, but understand *why*,
- uniqueness of ideas,
- elegance of ideas and
- for an academic engineer, publications.

Noticeably here, Jens privileges engineering over academe. 'Publish or perish' haunts academe, but even for an engineer working in a university, high skill levels related to engineering are the most significant prestige factor for Jens.

A-1: Joining A-Tech

I asked Jens why he had joined a spin-out, as distinct from a large company environment at the end of his PhD:

> *It's not just a way of earning money; I had a job offer from Jaguar while I was doing my PhD, but it wasn't what I wanted. The company* [USO/A-Tech] *was the best technology available in the jobs I was offered; in any case, a spin-out company is a good CV feature*

Here, Jens clearly privileges the reputation of a career engineer over immediate financial gain. For a mainstream academic, publication and tenure are at the heart of the career decision for exodus or stasis. For a newly qualified PhD engineer, the concerns are to be associated with leading edge technology, and a USO is a highly legitimate domain to work in in terms of supporting a future career; again, a different set of priorities to the career academic.

A-2: Working for A-Tech

Jens enjoys working for A-Tech as an engineer and at the outset holds a view that if a product is good enough it will sell. During A-2, however, he increasingly realises that this was a little naïve. During this period, the company acquires UK government funding in the form of two SMART awards to bring the technology to market. Although Jens was still very technology focussed, as he became more involved with the SMART work, he began to realise that the company's marketing strategy was weak and reactive. Operational business activities were undertaken by the company's founder, but Jens saw these activities as just having 'nuisance value' rather than any strategic dimension. At this time, although he was meeting sales people, he maintained what he referred to as 'a healthy amount of prejudice' against sales people, who were not valued, he explained, in engineering cultures if they had little or poor understandings of the product. He makes a clear distinction between operational business activities, which are seen just as a chore and the potential for growth through strategic marketing of an innovative product. This linkage between growth, innovation, strategy and company success can be seen as the trigger point where Jens develops a sense of professional self that is not solely defined through technological expertise, but also through the attractions of turning technology into market growth: a market pull, not just a technological push. In other words, the entrepreneurial side of business is not unattractive and indeed essential to success.

A-3: Developing through A-Tech and Education

Jens still enjoys working for A-Tech but is frustrated at the slow pace of growth and the lack of a strategic marketing strategy (perhaps, A-tech is unable to progress through the next 'critical juncture' to achieve the next phase of growth). In addition, he is still receiving a wage and does not have the prospect of equity. He realises the importance of business knowledge, but during this Masters education phase (as well as the concomitant business development activity), he also now realises this knowledge is not just about business functions but is on a higher level — it is about engaging with different vocabularies and audiences and about the appropriate presentation of self in new settings. At this point, I asked him if it was about networking. He explained that his early experiences of networking were poor 'not a valuable use of time, meeting people selling me low-level services such as business cards'. By September 2005, however, Jens had moved on from hovering at the edges of generic business networking events to building his image by giving a presentation at the Institute of Directors, a prestigious grouping of senior managers in the

United Kingdom. In his own analysis, he had gone from meaningless networking, where he was being sold things he did not want, to establishing a meaningful network of the right high-tech focussed contacts for his intended new business. Here, Jens highlighted the importance of being seen as a 'product developer' by potential customers, able to 'ask the customer, what is his pain, and how can my machine fix it'. To be seen as a student, or a recent student, or even an academic, was seen as a route to not being taken seriously. While he presents himself as a top-ranking engineer, this now seems to be enriched by a growing sense of entrepreneurial self.

B-1: The New Company

Now that Jens is in the process of setting up his own company, to pursue his expertise in the field under his own control, I asked him what he thought of 'entrepreneurs': had he 'become an entrepreneur'? Interestingly, he was fairly neutral towards the term, as he claimed it was not really used in Germany. However, he was developing a more positive view of the term as it was now associated with 'growth' as distinct from the mundanities of business life and the downside of his perception of sales. He was now beginning to see it as an attractive term, associated with growth and success. However,

> *it's not me yet – I'm still an engineer.*

I then asked him who his role models were: 'early stage developers in high tech spin outs, putting good technology ideas into practice'.

Here, Jens seems to be in the process of developing an enriched sense of engineering self-hood, with the elegance of solutions being extended from the laboratory into the market place. This manifest in three ways:

- He is focussed not just on recognising the value of business knowledge but also on presenting the 'right' identity in business settings.
- Most importantly of all is his maintenance of self as 'cutting edge engineer' and that spin-out activity is a legitimate dimension of engineering activity.
- Engineering identity is not compromised by association with growth and market share.

B-2: Out on His Own

Since 2005, B-Tech has been part of the university incubator in terms of registration and support but is located on Science Park premises, which are better able to meet his technical requirements. Jens is now focussed on building the business through developing customer relationships around a prototype developed with the R&D grant; the product concept has not changed significantly during this period, although it has been improved in terms of speed and efficiency. He is also seeking finance from business angels. I asked him whether his 'engineering self' still had the same importance in his company dealings as it had previously. He now considers that he has 'a totally different mindset' from when he first became involved in taking products to market. Though he is 'still an engineer', first and foremost, this is no longer defined primarily as technical expertise. He now presents himself as an effective and efficient solver of customer problems: as he achieves this through interactions with product developers and designers and quality control specialists in potential customer companies, his task is now to establish trust, 'engineer to engineer'. He crafts such legitimacy through making known his own track record, defined in terms of entrepreneurial and engineering acumen: he has, since 1998, designed instruments that have been used in laboratory settings and that have been sold to customers. He continues to publish in peer-reviewed conferences and journals and in trade journals. Having received SMART/R&D awards is a significant positive in terms of 'badging'. He makes less of the association with the incubators and the science park, as this can give the impression of being a little 'new'. In terms of understanding and developing business niche, he considers he has learned a tremendous amount over the past few years, but there is always something new to learn. The last thing he said during our final interview was that he 'still thinks as an engineer'.

Discussion

In this brief overview of Jens's transition through four phases of spin-out activity that tally strongly with the norms identified by Vohora et al. (2004), it is clear that Jens is carrying out considerable work on his identity. In line with Lounsbury and Glynn (2001), the self-hood is strongly associated with the purposeful development of legitimacy for himself and for the company, to achieve market share and growth, and that legitimacy is engineering-driven. As he makes his journey from PhD engineer in the research lab to founder of his own company, he actively crafts his career, his company and his identity in line with role expectations in his professional milieu (Goffman, 1959; Baruch, 2004; Down, 2006). He views spin-out/new technology development as high prestige work that may result in a high-reward firm eventually, or, if this is eventually unsuccessful, the experience will be seen as valuable on his CV by potential employers. At present, he is not considering an academic future but values academic recognition as playing a valuable role in validating his engineering identity. Thus far, no surprises. He has made the 'exodus' transition and is slowly but surely working his way through Vohora et al.'s critical junctures.

Yet, in approaching this case from the point of view of entrepreneurial identity, the case yields much more interesting insights that would not have been revealed otherwise. The most interesting outcome of this case is the extent to which Jens has wrestled with and resisted the notion of entrepreneurial identity. Initially, during his period with A-Tech, he discounts the business dealings as a necessary evil to avoid where possible. Later, he is drawn into taking on a more entrepreneurial role when he realises that the absence of strategic input in this area will likely compromise the possibility of success. Though he has recognised and moved on from his early naivety, for some time, he sees entrepreneurial identity as almost incompatible with being an engineer, an either/or vision where he is in danger of surrendering his engineering identity and ethos. Yet, towards the end of the A-Tech period and certainly during B-Tech, he has recognised that an entrepreneurial identity is a vehicle for many things; he begins to rework entrepreneurial identity as a mechanism to excel in his own style of engineering excellence as a high technology product developer. The transition is from professional engineer to engineering entrepreneur.

Conclusion

To conclude, the message of this chapter for theorists of entrepreneurial identity is that this case shows a powerful sense of agency and an empowered version of entrepreneurial identity. In that sense, the chapter contributes to the literature that suggests entrepreneurial identity can be crafted actively and used to strategic advantage (Lounsbury and Glynn, 2001) and furthermore that this can be achieved by entrepreneurs operating in the high technology domain. This is a novel contribution that adds to previous work, but what is more significant, and adds a layer of depth to the contribution, is that Jens does not move seamlessly from engineer to entrepreneur, casting off one set of identity clothes for another. A complex reworking of what it means to be an entrepreneur and what it means to be an engineer takes place that is enormously significant in his progression through Vohora et al.'s (2004) critical junctures. During his transition, he plays with Goffman's stereotypes, renders them fluid and puts them back together in his own vision of a powerful career trajectory (Baruch, 2004) where excellence is not compromised.

Practically, the career transition for either a mainstream academic (such as the founder of A-tech, though this in not the focus of this chapter) or a newly qualified PhD student is complex and challenging: but better understanding of the centrality of the engineering identity and its enrichment through entrepreneurial practice may well aid those engaged in the support of USO activity. Clearly, this is just one case, and more studies need to be carried out to explore these ideas in more depth.

Acknowledgements

Thanks to Jens (not his real name) for taking part in this study. Any errors or misinterpretations are mine entirely. Earlier versions of this chapter were presented at the High Technology Small Firms conference in 2006 at the University of Twente and to the KITE centre at Newcastle University Business School in 2007; thanks to participants in those events for many useful comments.

References

Anderson, A. R., & Chorev, S. (2006). Engineers learning to become entrepreneurs; stimulations and barriers in Israel. *International Journal of Continuing Engineering Education and Lifelong Learning*, *16*(5), 321–340.

Audretsch, D. B., & Erdem, D. K. (2004). *Determinants of scientist entrepreneurship: An integrated research agenda.* Discussion Paper 06/04. Max Planck Institute, Jena, Germany.

Baruch, Y. (2004). Transforming careers: From linear to multidirectional career paths: Organizational and individual perspectives. *Career Development International*, *9*(1), 58–73.

Bray, M. J., & Lee, J. N. (2000). University revenues from technology transfer: Licensing fees vs. equity positions. *Journal of Business Venturing*, *15*(5–6), 385–392.

Casson, M. (1995). *Entrepreneurship and business culture: Studies in the economics of trust* (Vol. 1). Cheltenham, UK: Edward Elgar Publishing.

Chell, E., & Allman, K. (2003). Mapping the motivations and intentions of technology oriented entrepreneurs. *R&D Management*, *33*, 117–134.

Cohen, L., & Musson, G. (2000). Entrepreneurial identities: Reflections from two case studies. *Organization*, *7*(1), 31–48.

Djokovic, D., & Souitaris, V. (2008). Spinouts from academic institutions: A literature review with suggestions for further research. *Journal of Technology Transfer*, *33*(3), 225–247.

Down, S. (2006). *Narratives of enterprise: Crafting entrepreneurial self-identity in a small firm.* Cheltenham: Edward Elgar.

Down, S., & Reveley, J. (2004). Generational encounters and the social formation of entrepreneurial identity – "young guns" and "old farts". *Organization*, *11*(2), 233–250.

Downing, S. (2005). The social construction of entrepreneurship: Narrative and dramatic processes in the coproduction of organizations and identities. *Entrepreneurship Theory and Practice*, *29*(2), 185–204.

Druilhe, C., & Garnsey, E. (2004). Do academic spin-outs differ and does it matter? *Journal of Technology Transfer*, *29*, 269–285.

du Gay, P. (1996). *Consumption and identity at work.* London: Sage.

Eisenhardt, K. M. (1989). Building theories from case study research. *Academy of Management Review*, *14*, 488–511.

Erikson, T. (2002). Entrepreneurial capital: The emerging venture's most important asset and competitive advantage. *Journal of Business Venturing*, *17*, 275–290.

Giddens, A. (1991). *Modernity and self-identity: Self and society in the late modern age.* Cambridge: Polity Press.

Goffman, E. (1959). *The presentation of self in everyday life.* Harmondsworth: Pelican Books.

Jenkins, R. (1996). *Social identity.* London: Routledge.

Lam, A. (2007). Knowledge networks and careers: Academic scientists in industry-university links. *Journal of Management Studies*, *44*(6), 993–1016.

Lounsbury, M., & Glynn, M. A. (2001). Cultural entrepreneurship: Stories, legitimacy, and the acquisition of resources. *Strategic Management Journal*, *22*, 545–564.

Luthje, C., & Franke, N. (2003). The making of an entrepreneur: Testing a model of entrepreneurial intent among engineering students at MIT. *R&D Management*, *33*, 135–147.

Meyer, M. (2003). Academic entrepreneurs or entrepreneurial academics? Research based ventures and public support mechanisms. *R&D Management*, *33*, 107–115.
Nicolaou, N., & Birley, S. (2003a). Academic networks in a trichotomous categorisation of university spinouts. *Journal of Business Venturing*, *18*, 333–359.
Nicolaou, N., & Birley, S. (2003b). Social networks in organisational emergence: The university spinout phenomenon. *Management Science*, *49*(12), 1702–1725.
Pirnay, F., Surlemont, B., & Nlemvo, F. (2003). Towards a typology of university spin-offs. *Small Business Economics*, *21*(4), 355–369.
Politis, D. (2005). The process of entrepreneurial learning: A conceptual framework. *Entrepreneurship Theory and Practice* (July), 399–424.
Reveley, J., Down, S., & Taylor, S. (2004). Beyond the boundaries: An ethnographic analysis of spatially diffuse control in a small firm. *International Small Business Journal*, *22*(4), 349–367.
Ritchie, J. (1991). Enterprise cultures: A frame analysis. In: R. Burrows (Ed.), *Deciphering the enterprise culture*. London: Routledge.
Storey, D. J. (1994). *Understanding the small business sector*. London: Routledge.
Van de Ven, A. H. (1992). Suggestion for studying strategy process, a research note. *Strategic Management Journal*, *13*, 169–188.
Vohora, A., Lockett, A., & Wright, M. (2004). Critical junctures in the growth of university high-tech spinout companies. *Research Policy*, *33*, 147–175.
Warren, L., & Anderson, A. R. (2005). Michael O'Leary: Entrepreneurial fire scorching the landscape. In: C. Steyaert & D. Hjorth (Eds), *The politics and aesthetics of entrepreneurship*. Cheltenham: Edward Elgar.

Chapter 4

The Roles of Scientists in the Start-up of Academic Spin-off Companies in the Life Sciences in the Netherlands

Harmen Jousma and Victor Scholten

Introduction

Academic knowledge can be put to use in a commercial environment in several ways. One such mechanism to transfer knowledge to the market place is the start of a new, separate company, termed an academic spin-off company, with the aim to commercially develop and exploit the knowledge generated in the university (Fontes, 2003). In 1999, the Dutch Ministry of Economic affairs published a paper stating that the number of high-tech start-ups in the Netherlands lags behind compared to other EU countries and the United States. Subsequently, initiatives were started to stimulate commercial exploitation of knowledge generated within universities. A specific initiative by the Dutch government in the area of the Life Sciences was the so-called Biopartner programme. This was started in 2000 with the objective to enhance the business climate for start-ups in the Life Sciences and to realize 75 start-ups within 5 years (Dutch Ministry of Economic Affairs, 1999). Actions were directed toward increasing awareness, stimulating starters, establishing facilities like a seed fund and academic incubators, and promoting the commercialization of academic knowledge within universities. A few years later, the Technopartner program and the Valorization Grant were implemented with similar instruments aiming at scientists in universities (Dutch Ministry of Economic Affairs, 2003).

Common in these programmes is that they have (or seem to have) the aim to stimulate scientists to turn their science into business by becoming entrepreneurs and creating new ventures. However, there is some debate on the matter of the scientist as

New Technology Based Firms in the New Millennium, Volume VII
Edited by R. Oakey, A. Groen, G. Cook and P. van der Sijde

an entrepreneur. There is consensus that the scientist's commitment to a start-up company is often of vital importance (Murray, 2004; Corolleur, Carrere & Mangematin, 2004). The basic reason for this is that in most cases the opportunity is in an embryonic state (Jensen & Thursby, 2001). During start-up, the scientists' know how in connection with the more formalized knowledge (prototypes, protocols, patent applications) is the key resource of the business. Often, his know how is unique and cannot simply be replaced by that of other scientists (Murray, 2004). In addition, the input of the scientist is necessary to guarantee a constant flow of innovations resulting in the development of a product portfolio (Vohora, Wright, & Lockett, 2004). They are also important as providers of social capital through their academic networks (Murray, 2004), which often extend into research departments of industry. Clarysse and Moray (2004) support an active role for the scientists in managing a start-up. On the basis of a process study, they conclude that in contrast to what venture capital firms like to see, appointing an outside CEO is not always a wise thing to do. The authors state that it is important to give the technology inventors time to learn how to run the business, supported by a business coach who is not involved in the company. According to their opinion this will create the possibility for the scientists to become CEOs themselves, or otherwise that their experience aids the acceptance of a CEO appointed from outside into the management team. Similarly, Samsom and Gurdon (1993) point out that possible conflicts between scientists and appointed management may be harmful for the success of the start-up or even cause its failure.

Other studies have emphasized that start-ups may be better off if managed by outside business experts (Franklin, Wright & Lockett, 2001) and that the number of PhDs in management is inversely correlated with company success in terms of value creation (Deeds, DeCarolis & Coombs, 1999). Another concern is that involvement of scientists in start-ups might distract them from their primary duties such as teaching and conducting research (Campbell & Slaughter, 1999), in other words a university scientist becoming an entrepreneur is a waste of scientific talent. At the same time, for the social capital of the scientist, it is important that the scientist retains close ties with the university and, therefore, is better not leave his academic position (Murray, 2004).

Related to the discussion on the role of the scientist in start-ups, Radosevich (1995) introduced the concept of surrogate entrepreneur. A surrogate entrepreneur is an entrepreneur from outside the originating organization that takes on a leading role to seize an opportunity and establish a business based on the research findings and knowledge created at universities. In a study among 57 UK universities, Franklin et al. (2001) found that universities that generate the most start-ups have more favorable attitudes toward surrogate entrepreneurs. Although universities might prefer the surrogate entrepreneur, the authors indicate that both approaches may not be mutually exclusive and that there may be benefits from combining the relative advantage of both (Franklin et al., 2001). Roberts and Malone (1996) propose several models in which they discuss the role of the scientist, a surrogate entrepreneur, capital investors, and the technology transfer office during different phases of the spin-off formation process. These phases are characterized by a developmental sequence in which the spin-off pauses at various steps and requires the entrepreneur to take adaptive actions and to change its behavior and practices to proceed to the next stage (Roberts & Malone, 1996; Vohora et al., 2004).

However, our insights into the role of academic and surrogate entrepreneurs during the various phases of the spin-off formation process remain limited. Previous research has not rendered a systematic overview of roles that scientists can have during the formation process of the academic spin-off. This research contributes to the field of academic spin-offs by focusing on the various roles that scientists play and analyzes how these roles may develop during the phases of spin-off formation. The issue motivating this line of research is whether it might be possible to optimize the role of the scientist in academic spin off formation. In the absence of a general rule, there might be guidelines or insights that can help work towards an optimal situation on a case-by-case basis. And even without a clear correlation between context, role, and result, enhancing consciousness of the various ways in which the spin-off process can be managed may help prospective academic and surrogate entrepreneurs make better choices. In addition, understanding their role might benefit university technology transfer professionals, investors, and business professionals who get involved in the academic spin-off process as well.

This chapter is structured as follows. First, we elaborate on the roles that scientists can play in the start-up of recently founded academic spin-offs. Second, to guide our inductive research we follow the literature of stage-based models of new firm development. Third, we discuss the empirical results that are analyzed and discussed in the conclusion and discussion section.

Theory

Entrepreneurial Roles

Research has identified diverse entrepreneurial roles for individuals during the formation of a start-up. These roles are often based on the activities or interests that individuals have. Pinchot (1985) distinguishes entrepreneurial roles based on the strength of vision and the depth of action. These roles have an important influence on how individuals operate and behave within a firm. Also Morris and Kuratko (2002) identify critical differences between entrepreneurial roles such as the traditional manager, the entrepreneur, and the intrapreneur. They elaborate on each role and explain, for instance, the primary motives these individuals have and their attitude toward the environment and internal operations. Similarly, Miner (1996) identifies four different types of entrepreneurs that use different strategies and use different agency roles to obtain entrepreneurial success. Following earlier research, we identify four entrepreneurial roles. First, the expert idea generator is the inventor with a strong technical expertise and is keen on developing new ideas. The personal achiever is the second entrepreneurial role and is characterized by the classic entrepreneur who has a strong vision and will to succeed. Third is the champion or super-salesperson who has a strong personal goal and is successful at pushing new ideas through the organization and thereby leveraging endorsement for the new venture. The real manager is the fourth role and is focused on strong managerial skills with a desire to control and gain power. This person is focused on enhancing the efficiency of the venture through organizing the activities in a systematic way.

The Inventor Role The academic inventor plays a critical role in transforming the scientific finding into a commercial opportunity. Academic inventors contribute to the entrepreneurial firm with their human capital and social capital. The human capital of the academic inventor encompasses not only the education but also the explicit knowledge such as patents and publications and the more tacit knowledge components such as capabilities and know-how (Agrawal, 2006; Murray, 2004). The social capital of the academic inventor refers to the academic network of contacts that the inventor has built over the years when moving through different faculty positions. The network of the inventor covers academic as well as business contacts around a disciplinary field. The inventor's network can help obtaining key resources and information from other leading scientists in the field (Murray, 2004).

The Entrepreneurial Role The role of the entrepreneur is to pursue and to exploit the opportunity. Often this exploitation is done regardless of the resources currently controlled (Stevenson & Jarillo, 1990). According to Stevenson's conceptualization, the entrepreneur is driven by the perception of opportunity and has a relative short commitment to that opportunity. Furthermore, the entrepreneur has often a strong locus of control (Brockhaus, 1975), he wants to be in charge of the business activities and be able to take the necessary decisions. Work should be guided by personal goals and not by those of others. The entrepreneur wants to be independent and is willing to assume calculated risks if the potential rewards are satisfying (Smilor, 1997). As such the entrepreneur is a good risk manager who calculates the value of the opportunity and compares it to the opportunity costs.

The Champion Role To overcome the indifference and resistance that major technological change provokes, a champion is required to identify the idea and promote the idea actively and vigorously through informal networks and critical organizational stages (Schön, 1963; Tushman & Nadler, 1986). Researchers have identified different processes for champions (Day, 1994) and different roles for champions (Venkataraman et al., 1992, Greene, Brush & Hart, 1999; Markham, 2000; Howell & Higgins, 1990). In many of these studies, championing is associated with large established firms in which individuals undertake corporate political roles to support an innovation. They identify themselves with a new idea, try to find resources for it and protect it from termination (Markham, 2000). Although the champion theory usually focuses on promoting and supporting new ideas within existing organizations, the underlying activities these champions perform are comparable to those of new venture entrepreneurs.

The Managerial Role The manager is, in contrast to the entrepreneur, more resource oriented and analyses the opportunity first before taking action (Stevenson

& Jarillo, 1990). The manager prefers to have a list of resources that are controlled and likes to negotiate over the necessary strategic course and plan the existing resource base accordingly. As the resources base is getting larger and more complicated, the need for management systems increase. These management systems include the use of capital allocation systems, formal planning systems, certain incentive systems, and others, all aimed at maintaining control over the complex resource base and reducing the risks (Brown, Davidsson & Wiklund, 2001). In addition, the larger resource base provides the manager with authority and power, which is related to the compensation scheme.

The roles explained here may be all present in the academic spin-off. More interestingly is that they may be needed at different times in the spin-off creation process. Depending on the risk profiles, the sources of motivation and the managerial or entrepreneurial capabilities, the scientists can take one or more different roles in the process of the spin-off venturing (Murray, 2004).

Phases of Growth

Stage-based models have been recognized by researchers to understand the process of new firm development. The phases that new firms pass through provide meaningful insight into the entrepreneurial actions that are prominent during a particular stage. Stage-based models have been presented in a different number of phases, but mostly recognized is the five-stage model by Greiner (1972). The model emphasizes that each phase contains a relative calm period of growth that ends with a management crisis. Periods of evolution and revolution take turn in the development of the new firm. Evolution is characterized by longed periods of growth during which organizational changes are linked to the particular environmental conditions (Aldrich, 1999). Revolution is characterized by substantial turmoil, during which the start-up team might take adaptive actions. On the basis of the five-stage model, Vohora et al. (2004) identify five phases in spin-off creation, successively the research phase, the opportunity framing phase, the pre-organization phase, the re-orientation phase, and the sustainable returns phase. At the intersection between two phases, revolution that takes place and a critical juncture needs to be overcome before the spin-off can proceed. The critical junctures are opportunity recognition, entrepreneurial commitment, credibility, and sustainability.

The *first phase*, the research phase, is the creation of know-how and technologies that may become commercially attractive. This is often at the university or the research institute where academic inventors develop the valuable technological assets (Vohora et al., 2004). Recognizing the opportunity to commercialize and to exploit the potential value of the technological asset in a new company is an important activity. In addition to recognizing the commercial opportunity, to start a new spin-off, it is essential to recognize the opportunity to start. This involves knowledge and experience to understand potential customer needs and translate these into the technological finding to make it more valuable.

The opportunity framing phase is the *second phase* and focuses on the examination whether the recognized opportunity has sufficient value to further pursue commercialization. This involves screening the technology for validity and to conceptualize the opportunity from a commercial perspective. Once the evaluation is satisfactory, the spin-off will develop strategic plans and acquire resources necessary to start. Important is the commitment of the person or team to put time and energy to all the activities needed to transform the idea into practice.

During the *third phase*, the pre-organization phase, the business model is formulated, which has essential impact on the directed path and the eventual success of the spin-off. The strategic plans are implemented based on the resources that the firm needs and the market it will enter. To acquire the resources, it is essential for spin-off entrepreneurs to gain the commitment of key individuals. The social capital of the entrepreneur is then essential to access and to secure resources and expertise and moreover build credibility and recognition for the company.

Having gained the credibility to access and acquire the resources, the spin-off can start to generate returns by commercially exploiting the academic invention. During that initial commercial exploitation, the entrepreneurial team faces challenges to continuously identify and integrate resources and then re-configure them. This *fourth phase* is the re-orientation phase during which the academic spin-off tests the size of particular markets, or the effectiveness of certain strategies (Cooper, 2001), and change or adapt accordingly. This may involve rather large changes in plans that had been formulated earlier due to information received from customers, competitors, or potential investors.

If the academic spin-off manages to reconfigure the business model and adjust it to internal and external changes, it may obtain sustainable returns, which is the last and *fifth phase*. The organization of the spin-off becomes more mature and optimized and management control systems will be put in place to coordinate the stock of resources and external relationships to achieve appropriate returns.

The role of the scientists may change during the phases of development dependent on personal preferences, career stage, institutional barriers, professional norms, and incentives provided by the spin-off or the academic institution (Murray, 2004; Audretsch & Stephan, 1996).

Study Design

Data Gathering

To obtain data about the start-up process of academic spin-offs and the role of the scientist in this process we chose start-ups that spun off from a university and were (originally) based on knowledge generated within that university. We restricted this study to the Netherlands and the life sciences sector to make comparisons more reliable. All 17 spin-offs that were included in the study were founded in the last 10 years, 13 were started 5 or less years before. Since we asked the interviewees to reflect on the start-up phase and draw on their memory, we only included relatively young companies to ensure the validity of the research. Also to improve the generalizability of the findings, we included spin-offs from all Dutch universities that have research programs in the field of the life sciences.

The interviews were held with the scientist who started a spin-off or was involved in the start-up process. Each interview was in-depth face-to-face and lasted for at least one hour. The aim was to construct overviews of the start-up process and clearly identify the role of the scientist during the spin-off formation. The interview was structured along four sections:

1. Spin-off: We asked about the core activities of the company, about the core technology, and whether major changes had been made in the business strategy.
2. Idea: We asked to reconstruct how the idea was recognized and which factors influenced the idea generation, for example, experiences from co-workers or a scouting policy of the university.
3. Market: We elaborated on the type of research that resulted in the spin-off, the scientist's market knowledge before starting the spin-off and how market knowledge was obtained.
4. Start-up process: This section contained questions on the exact contribution of the scientist to the business plan and in attracting finance. The scientists were asked about other persons who were important during the start-up process and how they contributed. We also asked whether the spin-off could have been started without the involvement of the scientist. Finally, we asked the subjects if on hindsight they would have done things differently and whether or not they had missed certain support during the start-up of the spin-off.

Sample

The academic spin-offs selected were all life sciences spin-offs. Of the 17 spin-offs, 14 were active in the biopharmaceutical domain, which involves identifying new drug targets, finding compounds to interact with new drug targets, developing lead compounds, and producing the drug. Also, three spin-offs were involved in diagnostics, vaccine development, and producing raw materials for drug development. Table 1 presents the characteristics of the spin-offs and the scientists.

Spin-off Characteristics As regards the type of business, we discriminated between products, services, and licensing. Spin-offs in life sciences often make a choice between developing a product and closing a licensing deal with a big industrial partner who will develop the product further. In some cases, the main technology of the spin-off can be used for multiple activities, for instance licensing or producing a product. Three spin-offs started without a patent, but all spin-offs obtained patent rights at a later stage. For the external investments, the source differed substantially. Venture capital was used in only five cases, whereas informal seed capital was obtained in seven cases, and research grants were also used in seven cases.

Table 1: Spin-off characteristics and scientist's background.

Spin-Off	Spin-off Characteristics					Scientist's Characteristics			
	Actual Type of Business	Patents at Start	Financing	Age	Nr of Employees	Full Time in Spin-Off	(Prior) Academic Position	Prior Industry Experience	Prior Industry Contacts
1	Products	Yes	VC	5	23		Professor		Yes
2	Licensing	Yes	Grants	5	2	Yes	Tenure	Yes	Yes
3	Service/products	Yes	VC	2	3		Professor		
4	Licensing	Yes	VC/Grants	4	5		Professor		Yes
5	Service/Licensing	Yes	Seed	1	2	Yes	Temporary		Yes
6	Service/products	Yes		3	2	Yes	Tenure		Yes
7	Service/products		Grants	7	16	Yes	Temporary		Yes
8	Products/licensing		Seed	10	6		Professor		
9	Service/products	Yes	VC	8	25		Professor		Yes
10	Products/licensing	Yes	Seed/ Grants	5	3		Professor		Yes
11	Products/licensing	Yes	Grants	5	3		Professor		
12	Products/service	Yes	VC, Seed, Grants	6	25		Professor		Yes
13	Products	Yes	Grants	1	1	Yes	Temporary		Yes
14	Products/licensing	Yes	Seed	4	3		Tenure		Yes
15	Products	Yes		2	2	Yes	Tenure		Yes
16	Service/licensing		Seed	3	3	Yes	Temporary		
17	Products	Yes	Seed	3	5		Professor		Yes

Scientist Characteristics Most of the scientists were full professors and in four cases it was a scientist without a permanent position (e.g., post-doc Ph.D. student). One of the scientists had industry experience, whereas all but four scientists had contacts with industrial parties before starting a business. These contacts usually originated from industry sponsored research projects or former licensing deals. Thirteen of 17 scientists were involved in applied research before the initiative to start a spin-off company. Also market knowledge resulted mostly from the interaction with business partners. For instance, if a scientist was involved in a certain disease area, the scientists became aware of the market size and the bottlenecks of treatments and clinical trials. This interaction with the market often provided the scientist with the knowledge about the potential of a certain technology and motivated him to start a new company. This does not necessarily mean that the scientist started the business himself, but the scientist could have taken the initial steps toward starting the spin-off. In four cases the initiative to starting a spin-off was taken by another person than the scientist.

Data

We analyzed the cases using a model for the development of spin-offs suggested by Vohora et al. (2004). This analysis provides us with a more detailed insight into the spin-off process and, more importantly, the role of the scientist is this process, which can vary among inventor role, entrepreneurial role, champion role, and managerial role during the phases of growth. The cases are subdivided into four development stages, which are opportunity framing, pre-organization, re-orientation, and the sustainability phase. The outcome of our investigation of the role of the scientist in these stages is presented in Table 2.

Empirical Evidence

The data suggest that the spin-off process is more differentiated than a distinction between academic and surrogate entrepreneurship can explain.

The analyses of the spin-off cases adds to earlier ones from process models of spin-off formation that only distinguish between the scientist as "just" the technologist or both the technologist and entrepreneur. The role of the scientist proved to vary significantly during the start-up process, and it was not always possible to apply a single role to the scientist in one single development phase. The role of the scientist fluctuated even within one development phase, indicating that it is not straightforward to characterize the role of the scientist. A clear example is the changing role of the scientist in the re-orientation phase in case 10. The company was facing bankruptcy and the scientist initiated a new business strategy, which resulted in the survival of the company. Similarly in case 12, the scientist temporarily had a bigger role in the management of the business when the company was going through a

Table 2: Analysis of the role of the scientist during the different growth phases.

Spin-Off	Opportunity Framing	Pre-Organization	Re-Orientation	Sustainable Returns
1	Scientist recognized opportunity and filed patent together with parent External agent pursued commercialization options	External person with industry experience starts spin-off Scientist almost not involved, but his experience was important to obtain credibility	Scientist is a scientific consultant and collaborates with the spin-off	Scientist is not involved. His university division is still a scientific partner to the spin-off
Scientist's role	*Inventor role*	*Inventor role*	*Inventor role*	*Inventor role*
2	Scientist works on opportunity, evaluates the potential, and files the patent	Scientist develops the spin-off further and performed extensive market research, obtained financing through a government grant, and wrote a business plan. Some experts were consulted	Scientist becomes CEO and develops two products that the company aims to license for production. The scientists decide to remain a research and development spin-off	The spin-off has not yet entered this phase
Scientist's role	*Inventor role* *Entrepreneurial role*	*Inventor role* *Entrepreneurial role* *Champion role*	*Inventor role* *Entrepreneurial role* *Management role*	
3	Scientist recognized various interesting applications. The scientist developed the technology	A VC company appointed a suitable CEO and provided capital. The CEO composed a business plan, sought finance	Scientist acts as CSO only. However, since the company heavily relies on fundamental research conducted at his academic division,	The spin-off has not yet entered this phase

	further and initiated a new spin-off	and established the company structure. Scientist was CSO and attracted potential clients through the scientific network	he is involved in determining the focus of the venture	
Scientist's role	*Inventor role* *Entrepreneurial role*	*Inventor role* *(Champion role)*	*Inventor role* *(Champion role)*	
4	Scientist was well experienced in both technology and market. Scientist recognized the commercial value, licensed technology to industry, and decided to initiate the spin-off	Scientist appointed a CEO who applied for government grants. Together they wrote a business plan and attracted VC money. Scientist is technological expert, and involved in strategic decisions	Scientist is CSO and focuses on creating new technologies and products Scientist is not involved on a business level. He only manages the company scientifically	The spin-off has not yet entered this phase
Scientist's role	*Inventor role* *Entrepreneurial role*	*Inventor role* *Managerial role*	*Inventor role*	
5	Scientists were involved in applied research, were close to the industry, and could recognize the technological needs and applications	Scientists decided to start spin-off and attracted grants, wrote business plan with feedback from a government programme. They started contacting possible clients who they already knew from their work at the university	Scientists ran company together and they could generate revenues from customers. Scientist identified that customers were more interested in specialists, accordingly they were approaching these customers and organized the spin-off	The spin-off has not yet entered this phase

Table 2: (*Continued*)

Spin-Off	Opportunity Framing	Pre-Organization	Re-Orientation	Sustainable Returns
Scientist's role	*Inventor role* *Entrepreneurial role*	*Inventor role* *Entrepreneurial role* *Champion role*	*Inventor role* *Champion role* *Management role*	
6	Scientist had close industry ties and conducted mostly contract research. Scientist spotted the opportunity and formulated the market value. Scientist was already entrepreneur in another company	Scientist decided to start spin-off. Scientist independently obtained grants and wrote business plan. Customers were acquired through his scientific network	Scientist became CEO and hired expertise in the area of business development	The spin-off has not yet entered this phase
Scientist's role	*Inventor role* *Entrepreneurial role*	*Inventor role* *Entrepreneurial role* *Champion role*	*Inventor role* *Entrepreneurial role*	
7	Scientists noticed the need for applied research. Their industry relations and experience helped formulating the research demands	Scientists decided to start providing the research service in a new company Using feedback from their network contacts and a government programme, they wrote a business plan	The scientists became CEO and first aimed at government projects company and later refocused on their own research lines	Scientists still CEO. Spin-off grows using revenues. Research leads evolve into products. Scientist seeks industrial partners for development

Scientist's role	*Inventor role* *Entrepreneurial role*	*Inventor role* *Entrepreneurial role* *Champion role*	*Inventor role* *Champion role* *Entrepreneurial role* *Management role*	*Inventor role* *Champion role* *Entrepreneurial role* *Management role*
8	University department develops and sells diagnostics and decides to set-up a spin-off. Also technology was commercialized through contract research	Two scientists were going to run the spin-off. Initial capital was acquired through a university-linked seed fund. A business plan was made by the scientists. Scientists approached new clients	Scientists were managing the spin-off in their spare time. Business was smaller than projected and was going bankrupt. Scientists left and university employee took over. He changed strategy to develop products and ended contract research	Employee became CEO. Employee attracts business partners to ally with in the commercialization of the products. Scientists are linked to spin-off merely for occasional advice
Scientist's role	*Inventor role*	*Inventor role* *Entrepreneurial role* *Champion role*	*Inventor role*	*Inventor role*
9	Scientist identified the potential of the technology through talks with industry. On the basis of ongoing research projects, the scientists were able to formulate the business proposition	University official and scientist decide to start a spin-off. It is agreed that finance should come from launching customer. For this purpose, a business developer is attracted. Scientist remains in charge but business activities are delegated	Business developer secures finance from launching customer. Scientist is CEO, and both prepare business plan to secure VC money. Reorientation follows from advice by capital provider. Scientist becomes CSO, but still leads spin-off	The spin-off has grown gradually. The scientist is still responsible for managing the company, although formally is not the CEO

Table 2: (*Continued*)

Spin-Off	Opportunity Framing	Pre-Organization	Re-Orientation	Sustainable Returns
Scientist's role	*Inventor role* *Entrepreneurial role*	*Inventor role* *Entrepreneurial role*	*Inventor role* *Entrepreneurial role*	*Inventor role* *Entrepreneurial role*
10	Through the scientist's experience in licensing IP, the opportunity was clear. However, the scientist never thought of starting a new company	An acquaintance with industry experience wanted to start a company and convinced the scientist to be part of the initiative. Scientist conducted scientific activities. Acquaintance carried out business activities, became CEO, invested own money and attracted finance. The scientist was to be CSO	CEO focused on growth and broadened the activities. He could not attract enough capital to maintain the strategy and the company faced bankruptcy. Scientist took initiative to continue on a smaller scale with a research orientation. He appointed others to the business, but kept responsibility	Scientist is CSO and appointed a CEO to manage the business. Scientist has more control over business strategy, which is related to scientific leads. The company has developed some IP, which might be licensed to industrial parties for further development
Scientist's role	*Inventor role* *Entrepreneurial role*	*Inventor role*	*Inventor role* *Entrepreneurial role* *Managerial role*	*Inventor role*
11	Scientist's experience enabled to spot the right applications for the IP. Industrial	Scientist commited to start IP-based spin-off. University offered some money	Both scientist and university partner run the spin-off. University has a	Profits and research grants are sufficient to fund new products. In the future, the

	party was interested in participating in research projects that could be commercialized	to start the additional research needed. Industrial partner hesitated. Together with university, scientist decided to sell products that were already developed	minority share in spin-off. Scientist writes together with partner a business plan and an advisory board is installed. New grants are used to start-up in new research areas	R&D division is expected to generate revenues. The scientist and his partner run the business on quality basis
Scientist's role	*Inventor role* *Entrepreneurial role*	*Inventor role* *Entrepreneurial role* *Champion role*	*Inventor role* *Entrepreneurial role* *Managerial role*	*Inventor role* *Entrepreneurial role* *Managerial role*
12	Scientist developed a technology that still needed further research. Scientist identified it as promising, through his knowledge in the field	Scientist did not actively pursue starting a spin-off. This was started after a business angel (BA) was willing to invest. BA attracted a second investor Investors appointed a CEO, who wrote a business plan. Scientist did the scientific part and became CSO	CEO did not perform well and hurt business. A new management team was appointed, who developed a new strategy. Company faced cash problems. Scientist applied for research grants and the research service component of the company was increased to obtain cash	Research service is performing well and has grown considerably. New products are in the pipeline. The scientist is CSO but has more influence in business matters compared to the first management team
Scientist's role	*Inventor role* *Entrepreneurial role*	*Inventor role*	*Inventor role* *Entrepreneurial role* *Champion role*	*Inventor role* *Entrepreneurial role* *Managerial role*

Table 2: (*Continued*)

Spin-Off	Opportunity Framing	Pre-Organization	Re-Orientation	Sustainable Returns
13	Scientist got involved in ideas to solve a certain market pull. The opportunities could not be worked out in university. Scientist started preparatory experiments in his spare time, which he financed himself	Government grant stimulated the scientist to develop and commercialize the ideas. A business coach helped writing a business plan and obtaining a research grant to further develop his technology	The company has not quite reached this stage yet. Scientist cooperates closely with a potential customer in the development. The scientist aims to become CSO and leave the business to a CEO that still has to be appointed	The spin-off has not yet entered this phase
Scientist's role	*Inventor role* *Entrepreneurial role*	*Inventor role* *Entrepreneurial role* Champion role	*Inventor role* *Entrepreneurial role* *Champion role*	
14	Scientists work on a solution for a medical problem that was encountered in hospitals. Proof of principle was obtained for a promising solution and the scientists filed a patent	With university official, scientists tried to close a production deal, but they failed. Patent was obtained by people who had connections with university and had business experience. They attracted finance and wrote a business plan. Scientists became science advisors	An industrial party will probably invest in the technology, which allows for production. Company is split into a commercial and an R&D division. Currently focus is on commercializing first product. One scientist is CSO and others act as advisors	The spin-off has not yet entered this phase

Scientist's role	*Inventor role* *Entrepreneurial role*	*Inventor role*	*Inventor role*	
15	Scientist was already working for a spin-off that developed a diagnostic test. When the test was market ready and the spin-off was not intended to commercialize it, the scientist took the opportunity	Scientist applied for government grant and met a business partner. They started a spin-off. No additional capital was needed as revenues from the sales were sufficient. Scientist and partner improved business plan to obtain additional grants	Scientist refocused at validation for the medical market. Revenues were still from sales in scientific market. Together they worked the market. The partner is handling most of the business activities, but they run the company together	The spin-off has not yet entered this phase
Scientist's role	*Inventor role* *Entrepreneurial role*	*Inventor role* *Entrepreneurial role* *Champion role*	*Inventor role* *Entrepreneurial role* *Champion role*	
16	Scientist always wanted to start a company. He recognized large market demands in his field of expertise	Scientist applied for a starter-grant where he met a partner with business experience to help with business matters. They both wrote a business plan. Scientist worked mostly on the scientific plan	Scientist and partner obtained capital. Research resulted in new opportunities, which were patented and changed strategy. Scientist as CEO and partner as manager run both the spin-off	The spin-off has not yet entered this phase

Table 2: (*Continued*)

Spin-Off	Opportunity Framing	Pre-Organization	Re-Orientation	Sustainable Returns
Scientist's role	*Inventor role* *Entrepreneurial role*	*Inventor role* *Entrepreneurial role*	*Inventor role* *Entrepreneurial role* *Champion role*	
17	Scientist is experienced with commercialization. After discovery of invention, he immediately decided to patent it. Scientist identified the value and initiated attracting capital	Scientist initiated spin-off and attracted industrial parties for co-development. He appointed someone start-up experience to become CEO. Both closed deal for co-development. CEO obtained funding. Scientist focused on science provided credibility for investor confidence	Multiple investors have been found and the R&D is progressing well. CEO is leading and has refocused to clinical trials. Scientist is scientifically involved	The spin-off has not yet entered this phase
Scientist's role	*Inventor role* *Entrepreneurial role*	*Inventor role* *Entrepreneurial role*	*Inventor role*	

difficult phase. These cases illustrate that during re-orientation the scientist can be a constant factor and ensure company survival in difficult times.

Reviewing the cases, we abstracted four distinct types of roles for the scientist, based on the number and level of involvement in certain roles. These four types are described below.

Business Run by Scientist (BRBS): Scientists who fitted this profile independently generated the idea to turn their science into business and were also responsible for all the activities needed in the start up process. They basically were involved in all types of roles, and occasionally, often when lacking experience, they sought support from experts in the form of advice and feedback. Usually they are still active in the company as a managing director or CEO and are in charge of the entire process of starting and running the business.

Business Run with Others (BRWO): These scientists are comparable to BRBS scientists with respect to the idea to start a spin-off, but they found a partner who was more experienced in business-related areas. As such they were less involved in the managerial role. During the start-up process, the scientist was involved in many non-science-related activities for instance getting approval for the spin-off among new clients or investors. In that position, the scientist took the champion role by promoting the spin-off.

Business Delegated to Others (BDTO): The idea and initial activities, such as contacting potential customers, to start a spin-off was initiated by the scientist. During the early phase the scientist took the role of inventor and entrepreneur. However the scientist did not have the ambition or believed he lacked the appropriate skills to actually manage the spin-off. Therefore, the scientist sought a qualified CEO or CBO (Chief Business Officer) to take the role of manager and delegated business related matters. However, also after the start of the spin-off the scientist remains in charge or has a dominating influence on the course of activities.

Business Run by Others (BRBO): In this role, the scientist did not necessarily initiate the start of the spin-off. The scientist may have pointed out the idea, but the initiative to start the company was taken by a third person (e.g., an investor or serial entrepreneur). This third person might become the CEO or will find a qualified person to manage the spin-off. The scientist has no control and his involvement in the start-up process is limited to scientific advice or performing research in support of the company's development. Hence, the scientist is involved in the inventor role and, or occasionally as an entrepreneur. This means that all other activities, even the opportunity recognition, are executed by others and the scientist has limited influence on start-up or running of the business. In Table 3 we describe the spin-off cases regarding the roles we identified here.

Discussion and Conclusion

This study analyzed the role of the scientist during the development phase of academic spin-off companies in the life sciences in the Netherlands and aims to shed

Table 3: The scientist role in the cases during the spin-off development.

Case Number	Development Phase			
	Opportunity Framing	**Pre-Organization**	**Re-Orientation**	**Sustainable Returns**
1	BRBO	BRBO	BRBO	BRBO
2	BRBS	BRBS	BRBS	–
3	BRBS	BRBO → BRWO	BRBO/BRWO	–
4	BRBS	BRBS/BRBO	BRBO	–
5	BRBS	BRBS	BRBS	–
6	BRBS	BRBS	BRBS	–
7	BRBS	BRBS	BRBS	BRBS
8	BRBS	BRBS	BRBS → BRBO	BRBO
9	BRBS	BDTO	BDTO	BRWO
10	BRBS	BRBO	BRBO → BDTO	BDTO
11	BRBS	BRBS → BRWO	BRWO	BRWO
12	BRBS	BRBO	BRBO → BRWO	
13	BRBS	BRBS	BRBS	–
14	BRBS	BRBO	BRBO	–
15	BRBS	BRBS → BRWO	BRWO	–
16	BRBS	BRBS → BRWO	BRWO	–
17	BRBS	BRBS → BDTO	BRBO/BDTO	–

more light on the issue whether or not scientists should be actively involved in running a spin-off. Using the development phases and entrepreneurial roles we identified, based on the cases, four different types of roles for the scientist: BRBS, BRWO, BDTO, and BRBO. This study provides an in depth analysis of the scientist's role and focuses on the exact activities performed by the scientist in the start-up process. Analyzing the scientist role indicates that this role varies considerably between development phases and can even change within a development phase.

Although this study focused on identifying different roles and did not seek explanation for the role of the scientist, some observations on the links between the role and the spin-off characteristics can be made. For instance, a spin-off with a specific core R&D activity often correlated with a BRBS type scientist involved. Furthermore, when comparing these companies to others with specific R&D as the core activity but with BDTO or BRBO scientist involved, there are clear differences. First, specific R&D spin-offs started by BRBS scientists were mostly financed by research or government grants, whereas spin-offs where business experts were involved, for example, BDTO- and BRBO-type scientists, more often used VC capital or public financing as a source of funding. Another explanation for the different roles might be affected by the intellectual property or academic knowledge of the spin-offs. Not all initial business ideas for a certain technology, as identified by the scientist, are interesting or commercially attractive for business experts or VC providers and consequently the scientist had to pursue the start of the spin-off by himself. Yet another factor might also be caused by the scientist himself. He or she may never need additional finance or may not want to exchange financial involvement for equity shares. In several cases we studied, the BRBS scientists made no attempts to find external finance. Similarly, for scientists with industry relations, we found a positive influence on the involvement of scientists in their academic spin-offs. Around half the BRBS scientists in our study had close relationships with industrial parties through their previous work experience. These relationships with industrial parties might have stimulated them to start a spin-off based on their research.

A noteworthy observation, based on Table 3, is that the role of the scientist time during the pre-organization phase and subsequent phases did not vary much for BRBS, BRWO, and BDTO scientists. However, for BRBO scientists, the cases show considerable variation both in time and between cases. The fact that a BRBO scenario took place in these cases may be caused by two important milestones in time resembling the first two of four critical junctures in spin-off formation (Vohora et al., 2004). The first juncture is opportunity recognition and takes place at the beginning of the opportunity framing phase. For the cases with low involvement of the scientist, the spin-offs would not have passed the first critical juncture without the initiative from outside parties. This does not mean that the scientists did not see commercial opportunities for the technology, but never considered the commercialization route. Entrepreneurial commitment is the second juncture, which involves tying emotionally committed individuals with sufficient experience to the business. During this phase, the scientist's involvement is needed to ensure new innovations (Vohora et al., 2004), but also business experience is needed and the scientist may not be the best candidate. In the cases with a relatively low commitment of the scientist to pre-organize the spin-off, outside parties were needed to pass this critical junction. Scientists perceived themselves as inexperienced with respect to running a business and consequently were not willing to give up their academic position. Another striking observation is the increase in involvement of the scientist in the re-orientation phase in cases 10 and 12. These spin offs encountered serious problems and the scientists were actively involved in coping with these problems and taking necessary actions including change of management. As a result, the scientist became more involved in running the business after solving the problems compared to the earlier situation. This indicates that although the scientist is not actively involved in running the business, his commitment can be very important as a constant factor in the company when facing serious problems. These observations might also imply that over time the scientist has become more familiar with business related matters enabling him be more actively involved in running the business. In these cases, both scientists stated that they learned the importance of being involved in running the business to ensure the fine tuning of both the business and the research strategies.

This study is based on an exploratory analysis of the role of scientists during the formation process of academic spin-offs in the life sciences. Future research might investigate a larger number of life sciences spin-offs and validate the roles as identified in this study. In addition, a larger number of spin-offs allows for statistical analyses to investigate the relationship between the scientist role and performance outcomes. Similarly, future research may also include other research areas where spin-off companies emerge such as computer sciences or physics and study possible differences or similarities with respect to the role of the scientist in the formation of the spin-off. Also, in the discussion we spoke about the differences between specific R&D spin-offs initiated by BRBS scientists and the spin-offs started by business experts. These differences may be a result of the sources of finance or the business model. Future research may investigate to what extent these differences influence the performance of spin-off companies. For instance, Roberts and Malone (1996) showed that multi-founder spin-offs were usually more product and market oriented and also evolved more rapidly compared to single founder spin-offs.

Ultimately, we believe, along these lines a more thorough understanding of the roles of the scientists in relation to the type of business, the way it is financed, its growth, and its success in terms of exploitation of the opportunity, can provide tools for scientists, entrepreneurs, technology transfer officers and policy makers to optimize the spin-off formation process on a case-by-case basis.

References

Agrawal, A. (2006). Engaging the inventor: Exploring licensing strategies for university inventions and the role latent knowledge. *Strategic Management Journal, 27*, 63–79.

Aldrich, H. E. (1999). *Organizations evolving*. London: Sage Publications.

Audretsch, D. B., & Stephan, P. E. (1996). Company-scientist locational links: The case of biotechnology. *The American Economic Review, 86*(3), 641–652.

Brockhaus, R. H. (1975). I-E Locus of Control scores as predictors of entrepreneurial intentions. *Proceedings, Academy of Management*, 433–435.

Brown, T. E., Davidsson, P., & Wiklund, J. (2001). An operationalization of Stevenson's conceptualization of entrepreneurship as opportunity-based firm behavior. *Strategic Management Journal, 22*(10), 953–968.

Campbell, T. I., & Slaughter, S. (1999). Faculty and administrators' attitudes toward potential conflicts of interest, commitment, and equity in university-industry relationships. *Journal of Higher Education, 70*, 309–351.

Clarysse, B., & Moray, N. (2004). A process study of entrepreneurial team formation: the case of a research-based spin-off. *Journal of Business Venturing, 19*, 55–79.

Cooper, A. C. (2001). Networks, alliances and entrepreneurship. In: M. A. Hitt, R. D. Ireland, S. M. Camp & D. L. Sexton (Eds), *Strategic entrepreneurship: Creating a new integrated mindset*. Oxford, UK: Blackwell.

Corolleur, C. D. F., Carrere, M., & Mangematin, V. (2004). Turning scientific and technological human capital into economic capital: The experience of biotech start-ups in France. *Research Policy, 33*(4), 631–642.

Day, D. L. (1994). Raising radicals: Different processes for championing innovative corporate ventures. *Organization Science, 5*(2), 148–172.

Deeds, D. L., DeCarolis, D., & Coombs, J. (1999). Dynamic capabilities and new product development in high technology ventures: An empirical analysis of new biotechnology firms. *Journal of Business Venturing, 15*, 211–229.

Dutch Ministry of Economic Affairs. (1999). *The entrepreneurial society: More changes and less hindrance for entrepreneurship (in Dutch)*. The Hague: Ministry of Economic Affairs.

Dutch Ministry of Economic Affairs. (2003). *Entrepreneurship in the Netherlands, knowledge transfer: Developing high tech ventures*. The Hague: Ministry of Economic Affairs.

Fontes, M. (2003). The process of transformation of scientific and technological knowledge into economic value conducted by biotechnology spin-offs. *Technovation*, *25*(4), 339–347.

Franklin, S. J., Wright, M., & Lockett, A. (2001). Academic and surrogate entrepreneurs in university spin-out companies. *Journal of Technology Transfer*, *26*, 127–141.

Greene, P. G., Brush, G. C., & Hart, M. M. (1999). The corporate venture champion: A resource-based approach to role and process. *Entrepreneurship: Theory and Practice*, *23*, 103–122.

Greiner, L. E. (1972). Evolution and revolution as organizations grow. *Harvard Business Review*, (July–August).

Howell, J. M., & Higgins, C. A. (1990). The champions of technological innovation. *Administrative Science Quarterly*, *35*, 317–341.

Jensen, R., & Thursby, M. (2001). Proofs and prototypes for sale: The licensing of university inventions. *The American Economic Review*, *91*, 240–259.

Markham, S. K. (2000). Corporate championing and antagonism as forms of political behavior: An R&D perspective. *Organization Science*, *11*(4), 429–447.

Miner, J. B. (1996). *The four routes to entrepreneurial success*. San Fransisco: Berrett-Koehler Publisher Inc.

Morris, M. H., & Kuratko, D. F. (2002). *Corporate entrepreneurship. Entrepreneurial development within organizations*. Fort Worth, TX: Harcourt.

Murray, F. (2004). The role of academic inventors in entrepreneurial firms: Sharing the laboratory life. *Research Policy*, *33*, 643–659.

Pinchot, G. (1985). *Intrapreneuring*. New York: Harper & Row.

Radosevich, R. (1995). A model for entrepreneurial spin-offs from public technology sources. *International Journal of Technology Management*, *10*, 879–893.

Roberts, E. B., & Malone, D. E. (1996). Policies and structures for spinning off new companies from research and development organizations. *R&D Management*, *26*(1), 17–48.

Samsom, K. J., & Gurdon, M. A. (1993). University scientists as entrepreneurs: A special case of technology transfer and high-tech venturing. *Technovation*, *13*(2), 63–71.

Schön, D. (1963). Champions for radical new inventions. *Harvard Business Review*, *41*, 77–88.

Smilor, R. W. (1997). Entrepreneurship: Reflections on a subversive activity. *Journal of Business Venturing*, *12*(5), 341–346.

Stevenson, H. H., & Jarillo, J. C. (1990). A paradigm of entrepreneurship: Entrepreneurial management. *Strategic Management Journal*, *2*, 17–27.

Tushman, M., & Nadler, D. (1986). Organizing for innovation. *California Management Review*, *27*(3), 74–92.

Venkataraman, S., MacMillan, I., & McGrath, R. (1992). Progress in research on corporate venturing. In: D. Sexton (Ed.), *State of the art in entrepreneurship*. New York: Kent Publishing.

Vohora, A., Wright, M., & Lockett, A. (2004). Critical junctions in the development of university high-tech spinout companies. *Research Policy*, *33*, 147–175.

Chapter 5

The Role of Knowledge Intensive Business Service Firms in University Knowledge Commercialisation

Kari Laine

Background and Theories in Use

Universities have a new role in the commercialisation of knowledge (Etzkowitz, 1998). The new role began with science parks and increased collaboration in 1980s and, with other forms of commercialisation, broadened to licensing and spin-off creation in 1990s, which also involved students (Rasmussen, Moen, & Guldbransen, 2006). Commercialisation has led to a situation where a complex web of relations exists between higher education, spin-offs created by them and large firms. All together the progress has been important because the 'commercialisation of knowledge connects the higher education to the users of the knowledge' (Etzkowitz, 1998). The rise of the knowledge-based society also brings the creation of knowledge-intensive firms into focus. The aim of the chapter is to create more understanding how small technology-based Knowledge Intensive Business Service (KIBS) firms can have a new role in knowledge commercialisation. In this chapter, the innovation chain is considered as a continuum from basic research through applied research to product development and finally commercialisation. There still exists a 'valley of death' between research and commercialisation (Markham, 2002). Spin-offs are one means to cross it.

The innovation process has to be managed as an entity. Generally, the innovation process consists of searching, selecting and implementing, which are all closely connected to learning (Tidd, Bessant, & Pavitt, 2005). New business opportunities are based on co-creation of value and discontinuities, in many cases on a combination of many discontinuities (Prahalad & Ramaswamy, 2004; Hamel,

New Technology Based Firms in the New Millennium, Volume VII
Edited by R. Oakey, A. Groen, G. Cook and P. van der Sijde

2000; Tidd et al., 2005). Sources of discontinuities that create opportunities may be new emerging markets or technology, new created political rules, mature industries running out of road, new market behaviour of customers, unthinkable events, new created business models or new regulation (Bessant et al., 2005). Idea screening and concept development of new products and services are less fuzzy if customer interactions occur already in the beginning of the innovation process (Alam, 2005).

Innovation capability can be targeted in four ways: products, processes, positioning and paradigm. This 'innovation space' also gives a model for innovation project portfolio management (Francis & Bessant, 2005). Innovations can also be divided as incremental, radical, continuous and discontinuous innovations. Systemic innovation is a combination of these, like using new technology in a new process and simultaneously rethinking the role of organisation. Systemic innovation is many times required in adaptation of innovation to make innovation accepted (Saranummi, Kivisaari, Väyrynen, & Hyppö, 2005). Successful innovation also requires integration and management of the whole innovation chain (Tidd et al., 2005). Research suggests that the entrepreneurs should concentrate more on organisational and market innovativeness than on technological innovativeness (Gans & Stern, 2003).

Innovation is a social process. It involves people meeting and sharing ideas. There are several types of innovation network including a new product or process development consortium, sector forum, new technology development consortium, emerging standards, supply chain learning, a cluster and a topic network. Different types of networks can be used when targeting different types of innovations. Operating an innovation network is difficult. Success factors for innovation networks are the following: partners with wide range of disciplines, science partners like universities, and access to investors and proactive management. Some of the challenges are how to manage something which is not in total control, how to operate on system level, how to build trust and shared risk taking without contracts and how to avoid free riders and spillovers. Configuring innovation networks must be in balance with innovation targets. (Tidd et al., 2005).

Dominant logic limits the ability to see new business opportunities (Prahalad, 2004). Therefore, focus should be on next practices, not on best practices. Low cost experimentation is needed to show the true potential of ideas (Prahalad, 2004; Hamel, 2000). Fast learning and articulation of experiments are needed. Firms must look beyond borders of industries and look beyond geographic borders and see exciting discontinuities instead of disruptive changes. Discontinuities challenge the dominant logic (Schumpeter, 1950). Seeing forward, forecasting, is considered to be one of the main competencies of the strategic thinking (Major, Asch, & Cordey-Hayes, 2001; Major, 2003). There are two usual failures in forecasting: the change is estimated to be faster than what will happen in reality and at the same time the impacts of the change are underestimated (Mannermaa, 2004).

Peripheral vision is a part of learning. The process consists of the sequential parts: scoping, scanning, interpreting, acting and learning and adjusting. Learning and adjusting affect mental models. To improve interpretation, appropriate channels have to be created to share and interpret information internally and externally. Frequent and free dialogue should happen spontaneously. It requires culture of trust, respect and curiosity. It must also be noticed that sharing information is important (Day &

Schoemaker, 2004). Scanning of the environment must be active. Passive scanning tends to reinforce old beliefs because the information comes mostly from known sources. Successful entrepreneurship requires not only analytical but also creative and practical intelligence, which all together constitute successful intelligence (Sternberg, 2004). Knowing is not enough: the knowledge has to be turned to action (Pfeffer & Sutton, 1999). Prior knowledge affects the ability to recognise the value of new information, assimilate it and apply it commercially. This is called absorptive capacity.

Learning and problem solving are close to each other. 'Problem solving skills represent capacity to create new knowledge'. Capability to learn is also connected to R&D. Firms that have their own R&D are also better in absorbing external knowledge. Absorptive capacity is therefore a critical part of innovation capabilities. 'Ability to assimilate new knowledge is a function of the richness of the pre-existing knowledge structure' (Cohen & Levinthal, 1990). Combining innovation with learning also suggests informal modes of technology transfer instead of traditional transfer models (Siegel, Waldman, Atwater, & Link, 2004). In a regional context, the 'innovative milieu' refers to physical and socio-cultural proximity, which makes up 'glue' that binds organisations together (Camagni, 2003). Learning from customers by observing them consists of a primary feedback on which the new service and process development can be based (Cunningham, 1994).

The context in which the process is managed brings up three elements: the strategic context, the innovativeness of the organisation and the connections of the organisation with the key actors in the environment (Rothwell, 1992; Tidd et al., 2005). The implementation can be divided into several core processes and their enabling support processes (Chiesa, 2001; Chiesa, Coughlan, & Voss, 1996). The operating environment is becoming more and more dynamic. New knowledge, new technology and new players make the situation more complex and accelerate the change. Development of the new knowledge is a continuum. New knowledge is built on existing knowledge. The timing must be correct to ensure that the new knowledge can create competitive advantage (Chiesa, 2001). Open innovation means that internal and external channels for idea generation and exploitation are considered as equal (Chesbrough, 2003). Even innovation can be outsourced, but it must be managed (Quinn, 1999, 2000). New forms or networking are emerging to reach continuous innovation in communities of creation and in the form of collaborative entrepreneurship (Miles, Miles, & Snow, 2005, 2006). Value is embedded in experiences and value is co-created with customers in interaction. There emerges an additional requirement for value creation: the experience network (Prahalad, 2004). We are moving towards the fifth generation innovation process, the key aspects of which are integration, flexibility, networking and parallel information processing (Rothwell, 1994). Value is created in interaction with customers (Prahalad & Ramaswamy, 2004).

Innovation diffusion is important in addition to innovation creation. There are five characteristics that may explain the success of an innovation, at least in so far as they affect the rate at which the innovation is adopted: relative advantage, compatibility, complexity, trialability and observability. Relative advantage means that the innovation must offer an advantage compared to the status quo. Compatibility refers to previous experiments and current needs of potential users of the innovation. The more complex is the innovation the less likely it is to be adopted. Trialability means that there should be a possibility to try out the innovation without total commitment to it at once. There should also be visible results from the use of the innovation for the users and for those who are observing the use. There are two main challenges: to support early adopters and to win mainstream credibility (Rogers, 1983, 1995, 2003).

In the future of KIBS, innovations are embedded in social activities. There are many kinds of innovations besides radical technological innovations. Innovation is closely linked to learning. Tacit knowledge has an important role in innovation and innovation is a complex process. Innovation diffusion is important in addition to innovation creation. Innovation is a collective undertaking and networks are essential for it (Toivonen, 2004).

A Case Study of a KIBS Firm

A longitudinal case study of a KIBS firm was implemented. The research covered a 9-year period from the startup. The aim of the case study was to find out how the firm is connected to the value network of Satakunta University of Applied Sciences (SUAS), how it manages its own innovation process (Tidd et al., 2005), how is the value network (Allee, 2003) of the firm built and how the configuration of its innovation networks (Tidd et al., 2005) changed during the time the research covered. To connect the present to the future, a scenario analysis was done by using the soft system methodology (Checkland & Scholes, 1990). In the research, case study principles were applied (Eisenhardt, 1989; Yin, 1994).

The founder of the firm was a student of SUAS. The student participated in R&D projects with regional Small and Medium Size Enterprises (SMEs) during his studies. The entrepreneur also gained knowledge from a special field of knowledge during his Bachelor thesis while he made a power system analysis for a large regional company. The entrepreneur had rich prior knowledge from most of the essential fields of his industry such as project management, R&D, power system analysis, field of industry and code of practice in serving the field of industry. He did not have much knowledge about running a business in practice, but he was a second generation entrepreneur and had a very entrepreneurial way of thinking.

The firm was founded in 1997 by the student at the time of graduation. It was the first tenant firm of the incubator in the SUAS. There was no IP transferred to the firm in the start phase, but there was both explicit and tacit expert knowledge transferred. The entrepreneur did not make any market research before starting the enterprise. He was convinced that his mentor had a right vision about the future of the service. The firm was totally funded by the entrepreneur himself with the help of a bank loan. The firm was started as a limited company with a minimum amount of share capital. During the first operating years, the firm also did automation and electrical system design for its customers. This provided positive cash flow to develop the new power supply management services. The first power system analyses were done for a large regional company in 1999. Serving new customers with analytical services started in larger scale in year 2002.

The company grew every year in turnover over 30%, although there were two unprofitable years 2003–2004, and it grew finally 63% last year. The turnover is now about 400,000 euros and the firm has six employees. The growth was not based on a single factor, but selling existing services to new customers and simultaneously developing new services have created most of the growth. The growth of the Internet created totally new business opportunities for the firm. A whole set of new services was created. The firm has also found a development partner for this. The new partner was also a firm started in the same incubator by students at the SUAS. In the firm, one-third of the personnel contribute to the development of new services and technology and about 5% of the turnover is applied to new service development.

Like many small firms, this firm also had many factors that hindered the development. During the first years, working mainly with one major customer was hindering recognition of new business opportunity and searching for new customers. The entrepreneur finally decided to hire a new employee to serve this customer to have more time to develop the new service and meet new potential customers. Having more customers later made it much easier to recognise business opportunities in the context of power system analyses. As the entrepreneur put it, 'Every time I meet my customers in their real environment, I see a lot of new business opportunities'. The main customer cases were all international and they had projects all over the world. The firm also gained fast a lot of multicultural experience.

Over these years, the entrepreneur learned appropriate ways to market his services. In the beginning, the contracts the firm made were not optimised from its point of view. The contracts gave customers too much scope to make delays in the process. In this case, recognition of new opportunities changed clearly when the new software product was launched. The firm was able to have many new customer contacts with this product. The firm immediately started to recognise new opportunities with accelerated rate. New features were added to existing services, and a totally new way to produce services is under development. But still the entrepreneur needs to meet the customer in his real environment to see opportunities. This requires trust to be created. Some of the customers are also returning customers because they have changes in their power systems and they need to update the analysis done earlier. The entrepreneur says that renting office space and special laboratory equipment from SUAS was essential for his startup. Later also, the expertise support from the SUAS was important because of the complex nature of the service.

In a technology-based firm, correct timing is essential. The entrepreneur must be able to meet the right people at the right time and to have the courage to start the business with a high level of risk. Also, funding the development is a challenge. In this case, other services, like electrical and automation engineering, were used to create positive cash flow to finance development. During these projects, the firm also gained multicultural experience. External funding was not needed. The entrepreneur even worked in another firm for short periods to have personal income during the first two years. Flexible risk funding for starting KIBS would be useful.

There was a strong supporting role from the SUAS in the beginning and during the whole development path. The support was in management, strategic thinking and technology development. Proximity to higher education has helped significantly. The entrepreneur has been able to use students in research and development projects in his firm. Also, all the new employees graduated from the same place that he did. Expert support has been available during the whole development period.

The best customers act as a development partner for a small firm. Although at the moment there are no intellectual property rights owned by the firm, in future, this may be one of the key issues. In a small firm, whether knowledge-based or not, small things matter. The entrepreneur must stay focused all the time, but even so unexpected events can change the whole promising success story. The start of a spin-off may be based on regional needs, but the most promising firms will grow to be at least national level players.

A hundred new firms have been created already in the same incubator and they form a new cluster of knowledge-intensive firms to the region. Many of the firms are in close collaboration with existing firms in the region increasing the competitiveness of them and creating added value for their customers. Finland has a strong national policy to create spin-offs from higher education. However, in 1997, it was not common to motivate students to start their business during the studies at university. At that time, SUAS already had a policy of creating spin-offs started by students. This has been a success for the region and a positive image builder for SUAS. Innovative new firms can also collaborate in large research and development projects administered by SUAS. This is one way to upgrade their knowledge and make them more embedded in regional clusters. Figure 1 describes the value network of a typical large R&D project administrated by SUAS. Firms form their own networks for innovation and service development. In Figure 2, the value network of the case firm is described. It is a simplified model where only the most important value-adding transitions between the key actors are marked. There are several industrial partners for new service development. The profits from the regular customers are used to fund new service development.

Tidd et al. (2005) argue that all innovation processes have the same general phases: search, selection and implementation. In this case, the process begins with active searching. In the search, the entrepreneur was able to exploit external connections effectively. The firm has a flexible strategy with a vision of 'Total power system analysis'. All development efforts are leading to that direction. The strategy changed many times as new customer needs were detected or dominant changes in environment and unexpected events occurred. The implementation was effective from the very beginning of the firm because the entrepreneur had experience of R&D. The entrepreneur took time to evaluate and reflect on experiences. Innovativeness was built in the organisation by constantly making small improvements to services. Learning required a combination of knowledge from several sources including also tacit knowledge.

Tidd et al. (2005) noted that networks must be managed according to innovation goals. In the beginning, the firm was helping other tenants in the incubator. The firm was also introduced to local clusters by the mentor. The entrepreneur himself started strategic partnership with industrial clusters. The next step is an option for heterogeneous innovation networks depending on the success of the ongoing service and process development. There was a simultaneous use of all the networks so that

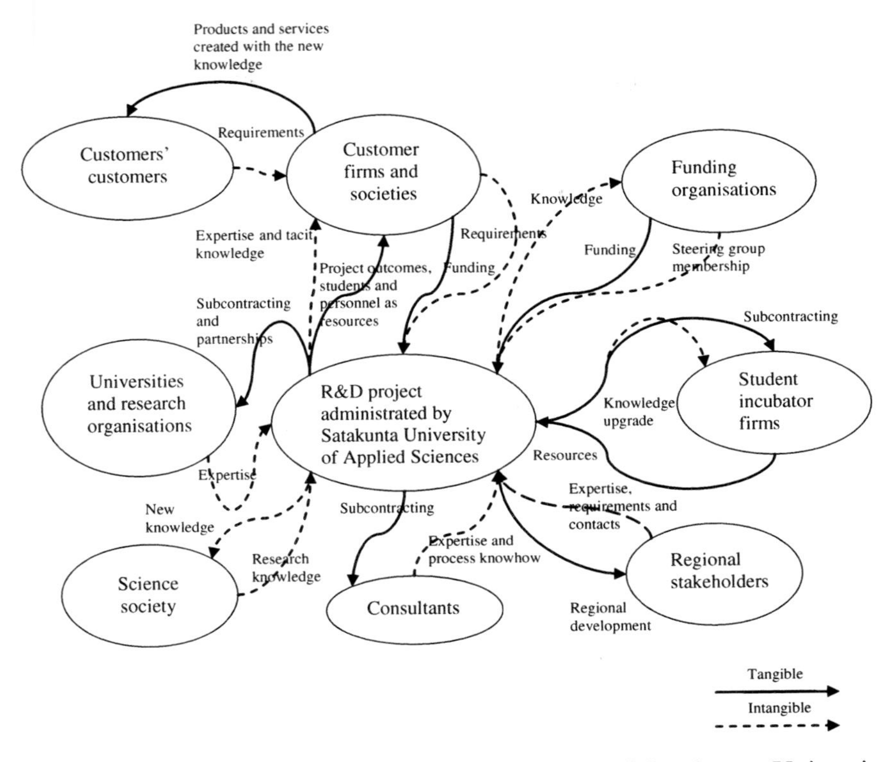

Figure 1: A typical value network of a R&D project of Satakunta University of Applied Sciences.

connecting to a new network did not mean the end of earlier network connections. In 2006, three future scenarios were built for the firm by scenario analysis using the soft system methodology. The analysis revealed that the entrepreneur has possibilities to choose between rather different development paths. Business as usual would lead to a strong position in Finnish markets but focusing on a specific area of services would give possibilities to grow international. Possibilities to exploit these new directions require success in ongoing development projects. It looks like there are much more opportunities than true possibilities to exploit them. Systemic innovation is one means to create possibilities to adjust the dimensions of innovation so that adoption comes more manageable.

Discussion and Implications from the Research

The entrepreneur had a rich prior knowledge structure and it helped him to have a high absorptive capacity. The absorptive capacity was further improved by R&D competence. Probably due to high absorptive capacity, the entrepreneur was able to

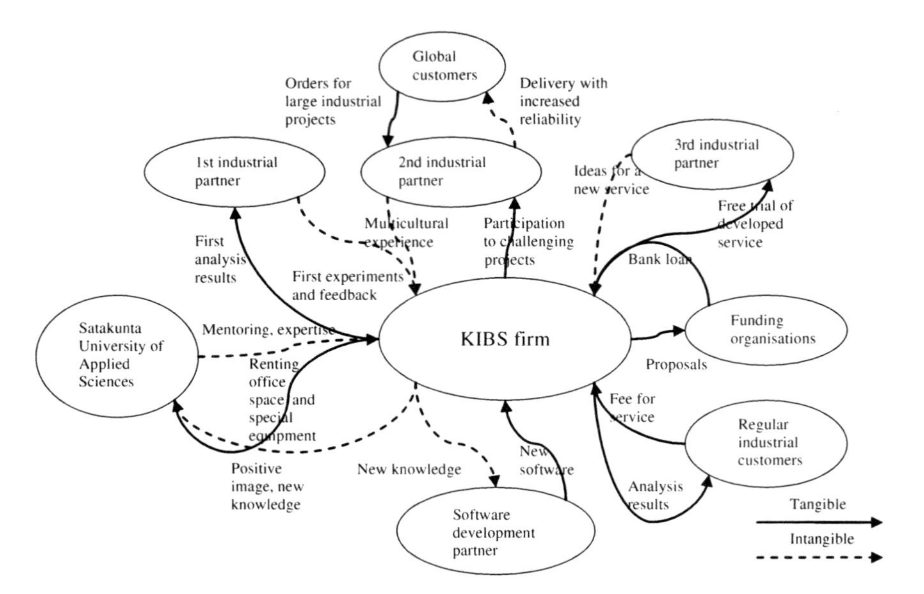

Figure 2: The value network of the case firm.

actively scan the business environment and technology knowledge. He was able to see and understand the forthcoming changes from weak signals and he was able to start new service development just in time. He was even able to create added value for the customers with new services that were based on the dominant changes. He was also able to challenge the dominant operating logic in the field. The firm also added to the absorption capacity of two local clusters by transferring knowledge to them.

The entrepreneur was able to transfer tacit components of knowledge in several occasions. He was also able to make low cost experimentations to prove the concepts he had created. He was able to manage the whole innovation chain and make implementation effective. Customers were connected to development projects in application phase, not too early or too late. The entrepreneur had a strategic approach to new service development. He was able to identify his weak points so that he could search for expert support in those fields. The entrepreneur sustained connections to the original source of the expert knowledge in SUAS all the time. The development steps have been small enough to ensure success in development, rapid service development and fast feedback from customers for double-loop learning.

The entrepreneur was not able to commercialise his knowledge alone, but he was able to create a value network to support service development and commercialisation. He was able to manage the value network without excessive contracts. He was able to prove the advantages of the service with support materials. The services were divided into easy-to-try parts to add to the trialability of the services and to accelerate the diffusion among potential users. Although the theory and process is complex, the service itself was easy to understand. The results of the service were observable. He was able to support early adopters and he earned credibility with mainstream users of the services through references from working applications. The entrepreneur was highly successful in building social networks. He embedded deeply in local clusters during the early phase of his entrepreneurship. He also had successful intelligence and concentrated more on markets than on technology itself. He used primary feedback from his customers to develop new services.

The implication from this case study for a small firm is the selection of partners. The customer partners could give insights into the industry and help in recognizing the new business opportunities. They can also co-develop the new services or products based on their needs, pilot the new services and offer a test platform for them.

The implication for regional development policy is that service development support for small KIBS when they commercialise knowledge from higher education is important. When supported, the KIBS firms could create many innovative services for regional clusters in traditional industries. These innovations are necessary for traditional industry to ensure its competitiveness. The support should only be used for product and service development and networking not for providing the services.

Implication to the higher education is to increase the embedding of small KIBS firms. Units of higher education can help in creating the regional embedding, in networking, in recognition of the technology trends and in scanning new research knowledge and in giving expert support. This includes creation of networks and actions that increase the interaction with regional clusters. They can also help in original opportunity recognition, in business planning and strategic thinking. The incubator should be able to resource a mentor for all tenants. However, every case is its own. Much depends on the mentor and most of all on the firm itself. The research also brings up the question of strategic and expert support. Renting special equipment lowers the threshold for small firms to launch new services based on the use of this sophisticated equipment. In many cases, an incubation process is essential to provide effective services for the entrepreneurs.

The KIBS firms are one channel to commercialise knowledge from higher education. Most of the commercialisation is still done by firms after collaboration projects when they use the transferred knowledge or technology in their new services or products. The KIBS firms are an important channel because they add the dynamics of the regional economy and offer opportunities of entrepreneurship for the students. The KIBS firms may also be interested to commercialise knowledge, which is not seen important by existing firms. For higher education, the KIBS firms create a live connection to users of knowledge when the interaction is further developed after the launch of the firm. The interaction also opens a whole spectrum of new opportunities for research in the field of collaborative innovation.

Acknowledgement

The researcher gratefully acknowledges funding from the High Technology Foundation of Satakunta and the Ulla Tuominen Foundation.

References

Alam, I. (2005). Removing the fuzziness from the fuzzy front-end of service innovations through customer interactions. *Industrial Marketing Management*, *35*, 468–480.

Allee, V. (2003). *The future of knowledge: Increasing prosperity through value networks*. Amsterdam: Butterwort-Heinemann.

Bessant, J., Lamming, R., Noke, H., & Phillips, W. (2005). Managing innovation beyond the steady state. *Technovation*, *25*, 1366–1376.

Camagni, R. (2003). *Regional clusters, regional competencies and regional competition*. Paper delivered at the International Conference on Cluster management in structural policy – International experiences and consequences for Northrhine-Westfalia, Duisburg, 5 December.

Checkland, P., & Scholes, J. (1990). *Soft system methodology in action*. Chichester: Wiley.

Chesbrough, H. (2003). *Open innovation: The new imperative for creating and profiting from technology*. Boston, MA: Harvard Business School Press.

Chiesa, V. (2001). *R&D strategy and organisation: Managing technical change in dynamic contexts*. London: Imperial College Press.

Chiesa, V., Coughlan, P., & Voss, C. A. (1996). Development of a technological innovation audit. *International Journal of Product Innovation Management*, *13*, 105–136.

Cohen, W., & Levinthal, D. (1990). Absorptive capacity: A new perspective on learning and innovation. *Administrative Science Quarterly*, *35*, 128–152.

Cunningham, I. (1994). *The wisdom of strategic learning: The self managed learning solution*. London: McGraw-Hill.

Day, G., & Schoemaker, P. (2004). Driving through the fog: Making at the edge. *Long Range Planning*, *37*, 127–142.

Eisenhardt, K. (1989). Building theories from case studies. *Academy of Management Review*, *14*, 532–550.

Etzkowitz, H. (1998). The norms of entrepreneurial science: Cognitive effects of the new university-industry linkages. *Research Policy*, *27*, 823–833.

Francis, D., & Bessant, J. (2005). Targeting innovation and implications for capability development. *Technovation*, *25*, 171–183.

Gans, J., & Stern, S. (2003). The product market and the market for "ideas": Commercialization strategies for technology entrepreneurs. *Research Policy*, *34*, 333–350.

Hamel, G. (2000). *Leading the revolution*. Boston, MA: Harvard Business School Press.

Major, E. (2003). Technology, transfer and innovation initiatives in strategic management. *Industry and Higher Education* (February), 21–27.

Major, E., Asch, D., & Cordey-Hayes, M. (2001). Foresight as a core competence. *Futures*, *33*, 91–107.

Mannermaa, M. (2004). *Heikoista signaaleista vahva tulevaisuus. [A strong future from weak signals, in Finnish]*. Helsinki: WSOY.

Markham, S. (2002). Moving technologies from lab to market. *Research Technology Management* (November–December), 31–42.

Miles, R., Miles, G., & Snow, C. (2005). *Collaborative entrepreneurship: How communities of networked firms use continuous innovation to create economic wealth*. California: Stanford University Press.

Miles, R., Miles, G., & Snow, C. (2006). Collaborative entrepreneurship: A business model for continuous innovation. *Organizational Dynamics*, *35*, 1–11.

Pfeffer, J., & Sutton, R. (1999). Knowing "what" to do is not enough: Turning knowledge into action. *California Management Review*, *42*(Fall), 83–108.
Prahalad, C. (2004). The blinders of dominant logic. *Long Range Planning*, *37*, 171–179.
Prahalad, C., & Ramaswamy, V. (2004). *The future of competition: Co-creating unique value with customers*. Boston, MA: Harvard Business School Press.
Quinn, J. (1999). Strategic outsourcing: Leveraging knowledge capabilities. *Sloan Management Review* (Summer), 9–21.
Quinn, J. (2000). Outsourcing innovation: The new engine of growth. *Sloan Management Review* (Summer), 13–28.
Rasmussen, E., Moen, O., & Guldbransen, M. (2006). Initiatives to promote commercialization of universtity knowledge. *Technovation*, *26*, 518–533.
Rogers, E. (1983). *Diffusion of innovations* (3rd ed.). New York: Free Press.
Rogers, E. (1995). *Diffusion of innovations* (4th ed.). New York: Free Press.
Rogers, E. (2003). *Diffusion of innovations* (5th ed.). New York: Free Press.
Rothwell, R. (1992). Successful industrial innovation: Critical success factors for the 1990s. *R&D Management*, *22*, 221–239.
Rothwell, R. (1994). Towards the fifth-generation innovation process. *International Marketing Review*, *11*, 7–31.
Saranummi, N., Kivisaari, S., Väyrynen, E., & Hyppö, H. (2005). Terveydenhuollon uudistaminen: Systeemiset innovaatiot ja asiantuntijapalvelut muutoksen ajureina [Renewal of healthcare: Systemic innovations and expert services as drivers of change, in Finnish]. Teknologiakatsaus 180/2005. Tekes, Helsinki.
Schumpeter, J. (1950). *Capitalism, socialism and democracy* (3rd ed.). New York: Harper and Row.
Siegel, D., Waldman, D., Atwater, L., & Link, A. (2004). Toward a model of effective transfer of scientific knowledge from academicians to practitioners: Qualitative evidence from commercialization of university technologies. *Journal of Engineering and Technology Management*, *21*, 115–142.
Sternberg, R. (2004). Successful intelligence as a basis for entrepreneurship. *Journal of Business Venturing*, *19*, 189–201.
Tidd, J., Bessant, J., & Pavitt, K. (2005). *Managing innovation: Integrating technological, market and organizational change* (3rd ed.). Chicherster: Wiley.
Toivonen, M. (2004). *Expertise as business: Long-term development and future prospects of knowledge-intensive business services (KIBS)*. Doctoral dissertation series 2004/2, Helsinki University of Technology, Laboratory of Industrial Management, Espoo.
Yin, R. (1994). *Case study research – Design and methods*. Newbury Park, CA: Sage.

Chapter 6

The Evolutionary Business Valuation of Technology Transfer

Mirjam Leloux, Peter van der Sijde and Aard Groen

Technology transfer from university to industry involving the commercial exploitation of academic technological inventions lately has been the subject of much investigation. Various authors have written on the impact that economic, social and ethical technology transfers have produced on society (Bozeman, 2000; Bercovitz & Feldman, 2003; Sampat, 2006; Litan, 2007) and on the ramifications of this process on universities (e.g. the possible neglect of basic science and/or education) (Arvantis, Sydow, & Woerter, 2008). At least four formal ingredients of technology transfer exist as follows. First, the development of Intellectual Property Rights (IPRs) by scientists. Second, collaborative research defines and conducts R&D projects jointly undertaken between enterprises and scientific institutions, either on a bi-lateral or on a consortium basis. Third, technology transfer is facilitated by the licensing of patented technologies to industrial enterprises. Fourth, technology transfer is triggered by the formation of start-up technology-oriented enterprises by researchers from the science-base generated at research institutes (Debackere & Vleugels, 2005; D'Este, 2007; Lowe, 1993; Brown, Berry, & Goel, 1991).

According to Harmon et al. (1997), there are numerous models that describe the process of technology transfer as linear progression of formal steps, in terms of informal networking arrangements, or a combination of formal searching and informal search arrangements, to ensure successful transfer. It seems clear that the process of knowledge transfer between universities and industry occurs through multiple channels, such as personnel mobility, informal contacts, consulting relationships and joint research projects, and that patenting and spin-offs play a comparatively small part in this process. For example, it has been estimated that approximately 20% of invention disclosures are patented and about 10% of these patents are licensed to firms (Mowery & Ziedonis, 2002; D'Este, 2007; Etzkowitz, Webster, Gebhardt, & Terra, 2000; Leydesdorff, 2001).

New Technology Based Firms in the New Millennium, Volume VII
Edited by R. Oakey, A. Groen, G. Cook and P. van der Sijde

Although many universities seem to be quite successful in commercialising their academic inventions, the development of a generally effective exploitation policy is often lacking. According to Bozeman (2000), effective technology transfer depends on the transfer agent, transfer medium, transfer object, transfer recipient, demand and the environment. The selection of commercialisation strategies may therefore depend on technological criteria (i.e. evaluating inventions on scientific and technological grounds), market criteria (i.e. based on characteristics of the market place) and policy criteria (i.e. referring to a governmental agency's resources and goals) (Lowe, 1993, 2002). The involvement of a technology transfer office within a university may stimulate the commercial exploitation of academic inventions (e.g. Litan, 2007; DeBackere & Veugelers, 2005). However, we believe that a business perspective towards an academic invention is crucial in developing its exploitation strategy. This business perspective may be regarded as a balance between technological and commercial opportunities on the one hand and different types of risks on the other (e.g. developmental risks, market-related risks and uncertainties regarding intellectual property protection). In the search for the most optimal balance, the development of adequate methods for the assessment of the value of academic research becomes crucial.

In this chapter, we contribute to this discussion by reviewing approaches to the business valuation of new technologies. First, we review conventional business valuation models, and then we propose some novel methods, based on social system theory.

Conventional Business Valuation Methods

Conventional models for the business valuation of technology are usually financially oriented and only measure economic value. Several of these financially oriented approaches have been reviewed by Leloux and Groen (2007). Current monetary (financial) valuation methods for technology include cost-based methods, income-based methods and market-based methods (Martin, 1999; Goldheim, Slowinski, Joseph, Edward, & John, 2005).

Cost-based methods are predicated on the costs involved in creating and developing or replacing assets under consideration and may additionally incorporate the benefit of introducing the product to the marketplace earlier. These calculations typically result in a lump sum value of the subject technology. Such an approach might be useful for valuing embryonic basic technology for which market applications have not yet been defined, or for technology that is easy to replicate or to 'design around'. The most serious drawback of this approach is that it makes no allowance for future income and/or profits streams, market conditions, buyer/seller motivations, useful life and risks associated with receiving future economic benefits that might accrue from a patent. In addition, this approach necessitates accurate cost and depreciation data that, in practice, may often be difficult to obtain.

Market-based methods account for market conditions. A useful guide to the value of a patent may be the price paid for a similar intellectual property asset in a very recent commercial transaction. Market-based valuation methods may also be based on comparable royalty rates, which are generally applied in specific industry segments. Other approaches include the calculation of Price/Earnings (P/E) ratios. Factors to consider include the nature of assets transferred, the industry and products involved, agreement terms and other factors that may affect the agreed upon compensation. Although this rationale is reasonable, public information on price and comparability is usually not readily available. Additionally, unusual intellectual property or disruptive technology portfolios do not have comparators that can be used as guidelines and require much more investigation.

Income-based methods account for future value using conventional discounted cash flow (DCF) methods. Income-based calculations generate a present value based on an estimation of future cash flows (through licensing or through direct exploitation), including their timing, expectations about variations in the timing of such cash flows, the time value of money as well as the costs of uncertainty and risk. The duration and timing of the cash flow is determined by forecasting through using common market diffusion tools, the useful lifetime of the asset (i.e. either as its physical or service life), its statutory or legal life (i.e. as a patent), its economic life (i.e. the period of time during which the property is producing an adequate return) and/or its functional or technological life (i.e. the period after which the technology becomes commercialised but before which the technology becomes obsolete).

The business risk associated with the realisation of the stream of expected cash flows may be captured through the use of appropriate discount rates, which are industry-specific and/or linked to the development stage of a novel technology. This method is often used for new technologies when comparators are not available and the potential market is large. The major drawback of this approach is the possibility of error due to subjective estimation. The DCF approach also does not account for the different risks and uncertainties associated with a multistage cash flow and variable discount rates over the lifetime of a patent. Thus, the valuation of the patent would need to be split into several distinct phases. Therefore, several risk-adjusted DCF methods have been developed, such as *Monte Carlo methods* (i.e. sensitivity analysis), the end result being a frequency distribution of net present values (NPV). The real role of such simulations is to understand the way in which values vary with the parameters of the model constructed. *Decision Tree Analysis* (DTA) accounts for more flexibility, since in each branch an underlying DCF analysis is performed, starting with the final analysis and working backwards in time to give a present value. *Option Pricing Theory* (OPT) methods, such as the continuous time Black–Scholes (B–S) and option pricing models, account for the changing risk of future cash flows. The present value of the returns a company will receive from an investment is likened to the value of the share, subject to a call option. Option pricing methods have some advantages. First, they circumvent the relatively arbitrary matter of the appropriate discount rate by employing the measurable volatility of an asset's index value as an indicator of risk. In addition, such pricing models can circumvent the uncertainties of future cash flows by employing index values based on the present value of assets in the same class.

Several authors advocate the application of option-based valuation methods in strategic investments rather than DCF models only and have developed multiple scenario integrated models to complement DCF and option pricing techniques (Slater & Mohr, 2006; Boer, 2000; Angelis, 2002; Cassimon, Engelen, Thomassen, & van Wouwe, 2004; Kossovsky, Brandagee, & Giordan, 2004). In practice, such multiple scenario valuation methods are applied by various multinationals in the pharmaceutical, chemical and energy industries, as well as consulting companies.

A Novel Experiential Stepwise Model for Valuation of the Business Technology

The predictive financial scenarios of conventional business valuation models, such as described above, depend on the inputs into these models, which are merely based on rather rough estimations of the technology's expected added value in the marketplace. Furthermore, these models are used in a transaction mode that, as is shown in marketing research, leads to an underestimation of the value generated over time between the knowledge producer and the user (e.g. Håkansson c.s., 1987, 1989, 1995, 1996; Narver & Slater, 1990; Ulaga & Eggert, 2006). Additionally, a focus on economic value alone may underspecify the value of university–industry transactions. Therefore, we have developed an experiential stepwise model for the business valuation of technology (Table 1). The model starts with the invention, passes through a stage of business awareness creation and defining its business perspective, then enters the phase of business development and technology transfer after having elaborated on a process of business valuation and includes technological, financial and business points of view (Leloux & Groen, 2007).

Following this approach, a context for an optimised integrated technology and monetary valuation model that can be used for practical applications is created. It may serve as a starting point or merely help with more rational decision making. The approach (described in Figure 1) indicates the activities needed at every stage of the technology transfer process. However, it uses only limited dimensions such as technology, market and the risks for business valuation of academic inventions. These are cultural, strategic and economic dimensions. We believe that multiple dimensions may be relevant as well. Therefore, we have developed an alternative approach for business valuation.

Business Valuation on the Basis of Social System Theory

As the monetary and technology valuation model may remain limited, we have attempted to develop a novel approach to the business valuation of academic research outcomes (i.e. technological inventions), using a much broader scope, which

Table 1: Stepwise approach for business valuation of technology (Leloux & Groen, 2007).

Business Valuation Stage	Description of the Process	Actor	Output
I. Invention	Research process	Scientist	Laboratory journals, research reports, scientific papers and publications
Intrinsic valuation of technology	Technology assessment	Scientist, peer review (consultant)	Technology scoring models, intellectual property (patent) scoring models resulting in an intrinsic technology value of the invention
II. Business awareness	Patent application	Scientist, patent attorney (sponsor)	Patent application, product application areas
III. Business Perspective	Market research	Business manager, consultant (scientists)	Business scan, e.g. • business need, • market drivers, • value chain, • macro/micro-segmentation, • market size, • market growth, • competitors SWOT analysis
Monetary valuation of technology	Financial scenario modeling	Business manager, consultant	Cost-based, market-based, income-based (DCF), option-based and multiple integrated scenarios models resulting in a market-based monetary value of the invention
IV. Business valuation	Business valuation process	Business manager, consultant, scientist	Integration of the technological and monetary value models resulting in an integrated business value of the invention

Table 1: (*Continued*)

Business Valuation Stage	Description of the Process	Actor	Output
V. Business development	Business development process, licensing	Business manager, consultant, lawyer, customer/ developer	Product/process information bulletin, secrecy agreements, license agreement, co-development contract
VI. Technology transfer	Technology implementation process	Scientist, customer/ developer	Detailed product/process technical information

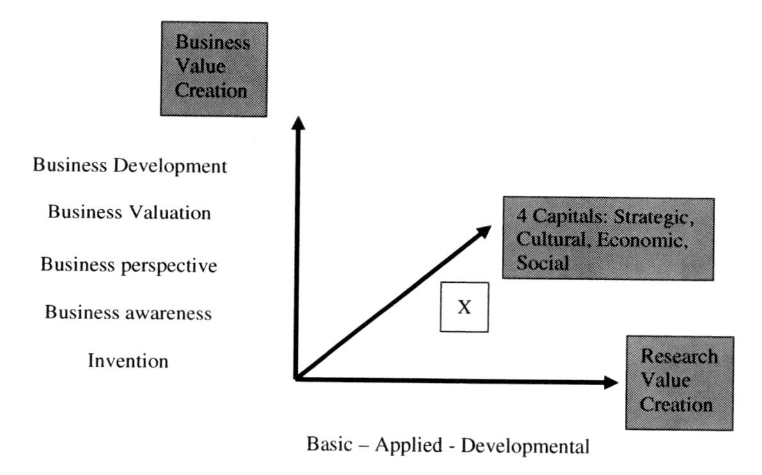

Figure 1: Multidimensional business valuation for technology.

is based on the entrepreneurial process theory, seen from a social systems theory perspective. In our model, the worth of an invention depends not only on economic value but also on the strategic, cultural or social value added by the developer of the invention, which is usually a company. Based on this social system theory approach (Groen et al. 2002a, 2002b; Rip & Groen, 2001), which is inspired by functional social system theory (i.e. Parsons, 1964), the inventor-entrepreneur (or entrepreneurial team) must develop four types of capital, namely strategic, economic, cultural (human and organisational), and social capital to a certain 'minimum' level (Groen, 2005) for a successful exploitation of an opportunity (Table 2).

The inventor-entrepreneur (or entrepreneurial team) needs to develop a strategy to develop knowledge and enter the market (*strategic capital*), identify and mobilise financial resources and develop a business model (*economic capital*) and find the

Table 2: Introducing the four-dimensional entrepreneurial process (Groen et al., 2002a, 2002b, 2005; Kirwan, van der Sijde, & Groen, 2006).

Strategic → value added to strategic position of developer
Cultural → value added to knowledge position of developer
Economic → value added to cost reduction or turnover of developer
Social → value added to network position of developer

necessary human and organisational resources (including the knowledge and technology resources) (i.e. *cultural capital*). Critical to all these activities is the network (*social capital*). Depending on the position of the inventor-entrepreneur (or entrepreneurial team) in the network, activities are channeled and facilitated but also constrained and inhibited (Dubini & Aldrich, 1991; Carsrud & Johnson, 1989; Hoang & Antoncic, 2003). Throughout the entrepreneurial process of research value creation, the inventor-entrepreneur (or entrepreneurial team) interacts with many actors in the network within and beyond the university, activating different parts of this network at different stages in the process to exchange and accumulate the capital and resources, power, knowledge, finance and networks necessary to pursue the opportunity.

We assume that the technology transfer of academic inventions may occur according to a value chain from basic science to applied science within the university context, moving towards the developmental research phase outside the academic context (i.e. in a developer company or spin-off firm outside the academic context). These three types of research within both an academic environment (basic and applied) and outside the academic environment in a spin-off firm (at the developmental research stage) can produce outcomes that may have utility for the developer, which produce better market relevance. Within the context of our social system theory involving opportunity recognition, opportunity preparation and opportunity exploitation, we can systematically analyse the value creation process using the four-dimensional model of entrepreneurial processes comprising strategic, cultural, economic and social capital (Table 2).

But what would then be the definition of 'value of a novel technological invention'? This value may be understood from the multidimensional viewpoint and the different perspectives involved, including the technology provider (usually the researcher) and the technology developer (i.e. the company). This multinational perspective is illustrated in Table 3. Usually, since (relatively) limited resources may be available at universities, the researcher may strive to increase access to economic capital, increased financial governmental and/or industrial funding for contract research collaborations, lump sum payments and royalties for licenses and money from investors in spin-off new ventures. Such financial resources may be more easily available from a company, which is willing to sponsor the research. From the viewpoint of a developer company interested in a novel technology, strategy may be more important since novel marketable products or services showing additional benefits to customers may be created, or even new product-market combinations may

Table 3: Examples of four capital dimensions in technology transfer (after Leloux, 2008).

Dimension	Technology Provider (University)	Technology Developing (Company)
Strategic capital: e.g. power, authority for goal attainment	Scientific leadership	Generation of new business/new product-market combinations Business leadership Profit maximization
Economic capital: relating to financial issues	(Relatively) limited resources available for scientific development Limited access to novel resources	Powerful resources available for development of profitable business Easy access to resources for novel opportunities
Cultural capital: relating to connections in cultural patterns and values or to knowledge necessary to influence patterns of behaviour (Bourdieu, 1979)	Knowledge of fundamental research, concepts, theories Idealistic, modest, not powerful	Knowledge of applied research, business knowledge Commercial, competitive, powerful, opportunistic
Social capital: relates to the network connections an actor can access, directly or indirectly (Burt, 1987; Burt & Celotto, 1992).	Scientific network	Business network (marketing & distribution)

be explored. For this developing company, investment in academic research will be expected to generate profit in the long run. Thus, for the developing company of the novel technology, 'value' will be different than for the academic inventor-entrepreneur.

Gross Business Value

To gain a deeper understanding of this 'value' concept, we have introduced the principle of *gross business value* in a former paper (Leloux & Groen, 2009). This *gross business value* may be defined as a function over time of the four dimensions of value: comprising economic, strategic, cultural and social. As a result, this *gross business value* of an academic technological invention can be positioned within a three-dimensional model, with the research value creation process on the x-axis, the stepwise business valuation process on the y-axis and its contribution to the developing firm's four types of capitals on the *z*-axis (Figure 1). The research value creation process describes a transformation from fundamental/basic science, through applied science, towards developmental science. Marketable products usually emerge from the developmental research phase. Fundamental/basic science usually occurs within academia, whereas developmental science occurs preferably at a developing company. The business value creation process describes a stepwise journey starting from an invention, through a phase of business awareness at the scientists, business perspective, business valuation and business development stages, which has been described in Table 1. And finally, the four dimensions of value are economic, strategic, social and cultural as indicated in Table 3.

Since this multidimensional approach contrasts with conventional business valuations of technology, how can we use this to define the value of an academic research outcome? Any novel academic invention could be subjected to this model. Because it would be situated within a research value creation process at the *x*-axis, it would also be located on the business value creation process (*y*-axis), and it would additionally add to different extents to the four capitals of the developer/company and/or provider/university (*z*-axis).

A Case Study

To enable a more profound understanding of this multidimensional business valuation tool, we conduct a demonstration by applying the model to a case study.

Recently, a technology license agreement between a Dutch university and a large pharmaceutical multinational has been mutually signed. The technology was a novel biotech production method for vaccines against *Streptococcus pneumoniae*. *Streptococcus pneumoniae* (Pneumococcus), which is a bacterial pathogen that leads to pneumonia, which, every year, is responsible for 500,000 cases in the United States. It leads to an estimated 40,000 deaths annually in the United States and an estimated 800,000 deaths yearly worldwide in children below the age of 5 years.

Approximately half of these deaths may potentially be prevented by vaccination. The global vaccine market is worth about 6 billions dollars a year, and it is expected to grow to over 11 billions by 2010.

Within the context of our novel multidimensional model (Figure 1), this invention (i.e. the novel vaccine production method) can be situated in the middle of the research value creation chain on the *x*-axis (i.e. applied science), indicating that considerable investment is still needed to bring it to the developmental research phase. In fact, this investment needs to be funded by the industrial developer of the invention, preferably a pharmaceutical company. Additionally, the invention may be situated somewhere halfway between the business value creation process at the *y*-axis, as awareness of a potential business potential gradually occurred to the academic inventor. We have indicated the positioning of this invention within both the business value chain and the research value chain as an 'X' in Figure 1.

We started the technology transfer process by contacting various pharmaceutical companies in the United States and Europe, potentially interested in Streptococci vaccines. But what would the value of this novel invention mean to these companies in view of our four types of capital? Some suggestions are indicated in Table 4.

Table 4: Multidimensional value creation for R&D collaboration/licensing, spin-off case and R&D collaboration.

Case	Capital	Developer/Company	Technology Provider-University
Licensing-R&D collaboration	Strategic capital	Increase (new product)	Increase (scientific leadership)
	Economic capital	Similar to increase (lower production costs, but also investment)	Increase (receives funding from external sources, i.e., the developer, and additionally royalty/ lump sum payments)
	Cultural capital	Increase (new knowledge)	Increase (new knowledge)
	Social capital	Increase (entrance to new scientific network)	Increase (entrance to new business network)
Spin-off (N.B. developer is spin-off)	Strategic capital	Increase (new product)	Decrease
	Economic capital	Increase (receives funding from venture capitalists; high profitability	Decrease
	Cultural capital	Increase (new knowledge)	Decrease due to spin-out
	Social capital	Increase (business network)	Decrease
R&D collaboration	Strategic capital	Limited effect	Increase (scientific leadership)
	Economic capital	Limited effect	Some increase (receives funding from external sources, i.e., the developer)
	Cultural capital	Increase: new knowledge	Increase: new knowledge
	Social capital	Increase: new scientific networks	Increase: new business networks

Upon adopting this novel vaccine production technology, pharmaceutical companies would be able to generate a new, patented Streptococci vaccine that could be produced (potentially) at low cost. This new business generation would thus add to their strategic capital, and also to their economic capital, as production costs could be lowered. However, investment also would be needed to develop this technology before any market entry. Their collaboration with the university would further add to a new scientific network (social capital) and new knowledge (cultural capital). The researchers at the university, on their part, were interested in setting up a licensing and R&D collaboration with the company, to serve both its interest in further scientific leadership (increase of strategic capital) increase of business-related network (social capital), new knowledge development (cultural capital) and financial resources (economic capital) through both funding of research and royalty/lump sum payments.

In contrast, however, the pharmaceutical companies might have perceived that the developmental risks and costs were too high and thus also might have decided not to enter into any form of a collaboration and/or licensing agreement with the university. In this alternative scenario, the researchers at the university could potentially also have decided to create a spin-off, to further develop the novel vaccine production technology. In our model, the formation of an academic (biotech) spin-off enhances economic capital (high investments by, e.g., venture capitalists, potentially also high profitability) and social/network capital (especially business networks). However, the effect on academic cultural capital implies limited new academic skills or knowledge development, since spin-offs usually lack the resources necessary to extensively fund R&D collaboration programs at universities.

For the university, the *gross value* of their invention depends thus on the type of chosen exploitation route between spin-off creation, R&D collaboration or licensing. An R&D collaboration between an academic centre and a company will have less economic impact on the university (i.e. low investments, medium profitability), but both partners will gain in social/networking capital (i.e. enjoyed by business and scientific networks) as well as in academic cultural capital (i.e. involving new academic skills or knowledge development) and company cultural capital (i.e. through new knowledge development). Based on our model, we believe that, for the university, the gross value of R&D collaboration and/or licensing is much higher than for spin-off development (Table 3).

For companies, the *gross value* of a novel technology depends on the balance between investments (development costs, time-to-market) and profitability (economic capital) and strategic capital (new business development). An academic invention may be a quite novel, even disruptive, radical technology, with an expected high profitability/investment ratio, that opens up unknown scientific and business networks for the company. Such an invention would be of much greater *gross business value* to the company than an alternate case, for example, a continuous incremental innovation, with a limited profitability/investment ratio and no access to relevant novel networks. For the developing companies of the novel technology, the positioning of the invention within the research value creation process is relevant as directly linked to time-to-market and investment costs.

Final Remarks

In this chapter, we have tried to review some general methods of business valuation of new technology. In addition, we have introduced our novel evolutionary approach to this field. We strongly believe that strategic commercial exploitation of academic inventions needs a generally applicable business valuation approach, going beyond conventional models. We believe that our multidimensional approach, based on social system theory, may open up new avenues. In a follow-up paper, we will go into further detail regarding the application of our model to case studies.

References

Angelis, D. I. (2002). An option model for R&D valuation. *International Journal of Technology Management*, *24*(1), 44–56.

Arvantis, S., Sydow, N., & Woerter, M. (2008). *Is there any impact of university-industry knowledge transfer on the performance of private enterprises? An empirical analysis based on Swiss firm data, Review of Industrial Organization*, 32(2), 77–94.

Bercovitz, J., & Feldman, M. (2003). Technology transfer and the academic department: Who participates and why? Paper presented at the DRUID Summer Conference 2003 on Creating, sharing and transferring knowledge. The role of geography, institutions and organizations, Copenhagen.

Boer, F. P. (2000). Valuation of technology using 'real options'. *Research-Technology Management*, *43*(4), 26–30. Industrial Research Institute, Inc.

Bourdieu, P. (1979). 3 stages of cultural capital. *Actes de la Recherches en Sciences Sociales*, *30*, 3–6.

Bozeman, B. (2000). Technology transfer and public policy: A review of research and theory. *Research Policy*, *29*, 627–655.

Brown, M. A., Berry, L. G., & Goel, R. K. (1991). Guidelines for successfully transferring government-sponsored innovations. *Research Policy*, *20*, 121–143.

Burt, R. S. (1987). Social contagion and innovation: Cohesion versus structural equivalence. *The American Journal of Sociology*, *92*(6), 1287–1335.

Burt, R. S., & Celotto, N. (1992). The network structure of management roles in a large matrix firm. *Evaluation and Program Planning*, *15*, 303–326.

Carsrud, A. L., & Johnson, R. W. (1989). Entrepreneurship: A social psychological perspective. *Entrepreneurship and Regional Development*, *1*(1), 21–31.

Cassimon, D., Engelen, P. J., Thomassen, L., & van Wouwe, M. (2004). The valuation of a NDA using a 6-fold compound option. *Research Policy*, *33*, 41–51.

Debackere, K., & Veugelers, R. (2005). The role of academic technology transfer organizations in improving industry science links. *Research Policy*, *34*(3), 321–342.

D'Este, P. (2007). University-industry linkages in the UK: What are the factors underlying the variety of interactions with industry? *Research Policy*, *36*(9), 1295–1313.

Dubini, P., & Aldrich, H. (1991). Personal and extended networks are central to the entrepreneurial process. *Journal of Business Venturing*, *6*(5), 305–313.

Etzkowitz, H., Webster, A., Gebhardt, C., & Terra, B. R. C. (2000). The future of the university and the university of the future: Evolution of ivory tower to entrepreneurial paradigm. *Research Policy*, *29*, 313–330.

Goldheim, D., Slowinski, G., Joseph, D., Edward, H., & John, T. (2005). Extracting value from intellectual assets. *Research-Technology Management*, *2*(March–April), 43–46. Industrial Research Institute, Inc.

Groen, A. J. (2005). Knowledge intensive entrepreneurship in networks: Towards a multi-level/multi dimensional approach. *Journal of Enterprising Culture*, *13*(1), 69–88.

Groen, A. J., During, W. E., & Weaver, K. M. (2002a). Alliances between HTSF's and their partners: A multidimensional process approach. In: R. Oakey, W. E. During & S. Kauser (Eds), *New technology-based firms in the new millennium* (Vol. 2, pp. 197–217). Oxford: Pergamon.

Groen, A. J., Weerd-Nederhoff, P. C., de Kerssens-van Dongelen, J. C., Badoux, R. A. J., & Olthuis, G. P. H. (2002b). Creating and justifying research and development value: Scope, scale, skill and social networking of R&D. *Creativity and Innovation Management*, *11*(1), 2–16.

Håkansson, H. (Ed.) (1987). *Industrial technological development: A network approach.* London: Croom Helm.

Håkansson, H. (1989). *Corporate technological behaviour; co-operation and networks*. London: Routledge.

Harmon, B., Ardishvili, A., Cardozo, R., Elder, T., Leuthold, J., Parshall, J., Raghian, M., & Smith, D. (1997). Mapping the university technology transfer process. *Journal of Business Venturing*, *12*, 423–434.

Hoang, H., & Antoncic, B. (2003). Network-based research in entrepreneurship, a critical review. *Journal of Business Venturing*, *18*(2), 165–187.

Kirwan, P., van der Sijde, P., & Groen, A. J. (2006). Assessing the needs of new technology based firms (NTBFs): An investigation among spin-off companies from six European Universities. *The International Entrepreneurship and Management Journal*, *2*, 173–187.

Kossovsky, B., Brandagee, B., & Giordan, J. C. (2004). Using the market to determine IP's fair market value. *Research-Technology Management*, *47*(3), 33–42. Industrial Research Institute, Inc.

Leloux, M. S. (2008). Negotiations in technology transfer. In: S. J. HillesØe (Ed.), *Negotiation, the art of reaching agreement*. Denmark: Academia.

Leloux, M. S., & Groen, A. J. (2007). Business valuation of technology: An experiential model. *Les Nouvelles* (September), 478–486.

Leloux, M. S., & Groen, A. J. (2009). Estimating business value of academic research outcomes: Towards a multi-dimensional approach. *International Journal of Technology Transfer and Commercialization*, *8*, 3–21.

Leydesdorff, L. (2001). The transformation of universityindustry-government relations. *Electronic Journal of Sociology*, 1–27.

Litan, R. E. (2007). *Commercializing university innovations: A better way*. NBER Working Paper, Vol. JEL No. 018, M13,033,034,038, pp. 1–34.

Lowe, J. (1993). Commercialisation of university research: A policy perspective. *Tehcnology Analysis and Strategic Management*, *5*(1), 27–37.

Lowe, R. A. (2002). *Entrepreneurship and information asymmetry: Theory and evidence from the University of California*. Unpublished Working Paper. Haas School of Business, U.C. Berkeley, Berkeley, CA.

Martin, M. (1999). Marketing of advanced materials intellectual property. Paper presented at the Twelfth International Conference on Composite Materials, July 8, Paris, France.

Mowery, D. C., & Ziedonis, A. A. (2002). Academic patent quality and quantity before and after the Bayh-Dole act in the United States. *Research Policy*, *31*, 399–418.

Narver, J. C., & Slater, S. F. (1990). The effect of a market orientation on business profitability. *Journal of Marketing*, *54*(4), 20–35.

Parsons, T. (1964). *The social system*. New York: The Free Press.
Rip, A., & Groen, A. J. (2001). *Many visible hands* (pp. 12–37.). Cheltenham: Edward Elgar.
Sampat, B. N. (2006). Patenting and US academic research in the 20th century: The world before and after Bayh-Dole. *Research Policy*, *35*, 772–789.
Slater, S. F., & Mohr, J. J. (2006). Successful development and commercialization of technological innovation: Insights based on strategy type. *Journal of Product Innovation Management*, *23*, 26–33.
Ulaga, W., & Eggert, A. (2006). Value-based differentiation in business relationships: Gaining and sustaining key supplier status. *Journal of Marketing*, *70*(1), 119–136.

Chapter 7

Post-Project Market Review as a Tool for Stimulating Commercialisation of Knowledge Creation Projects

Liesbeth Y. Bout, Jaap H. M. Lombaers,
Efthymios Constantinides and Petra C. de Weerd-Nederhof

Post-Project Reviews are mainly used as a tool to improve organisational learning (Busby, 1999; von Zedtwitz, 2002). However, the concept of post-project review can also be used as a tool to identify new market potential and to hand over technical knowledge from technical to marketing personnel (von Zedtwitz, 2002). This chapter presents the findings of a research project on the improvement of commercialisation at a research organisation. After the problem analysis, a session based on the concept of post-project reviews is introduced as one of the potential solutions to improve the commercialisation of knowledge creation projects.

Introduction

This chapter addresses the introduction of a post-project market review, which is based on the concept of post-project reviews to stimulate commercialisation. It will start with a brief description of the case-company. After this, the motives of the research will be clear, and the research methodology will be explained in Chapter 2.

Case-Company

The case-company is a publicly funded Dutch research organisation for applied scientific research. Its mission is to generate knowledge based on scientific research and develop applications with the aim of strengthening the innovative power of the industry and the public sector.

New Technology Based Firms in the New Millennium, Volume VII
Edited by R. Oakey, A. Groen, G. Cook and P. van der Sijde

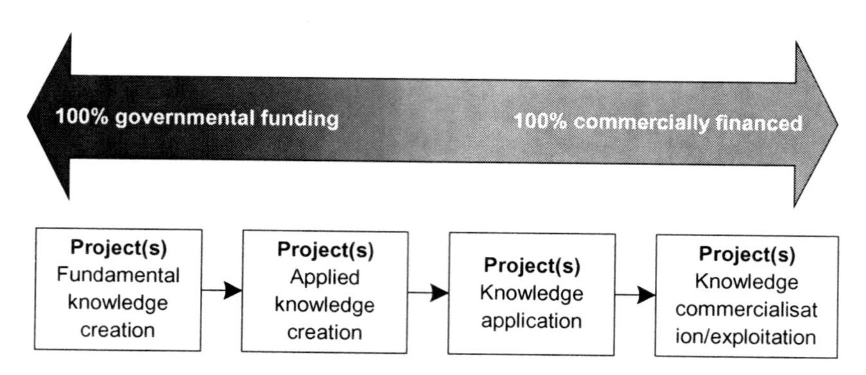

Figure 1: Innovation process and financing.

The government finances the research either wholly or in part and public funding makes up about 30% of the institute's budget. The fundamental knowledge-creating projects with governmental funding have a duration of one year, during which the financing of the project is guaranteed. After that, a follow-up project can be started for further development of the created knowledge and additional funding is possible.[1] If projects became more application oriented, they must be increasingly financed by the industry (Figure 1).

Problem Definition

Past experience indicates that a considerable part of the knowledge created in the more fundamental, government-funded projects remains unused by both government and industries. The consequence is that this knowledge cannot be used for creating value and gain commercial funding that would strengthen the innovative power of industry and government. The purpose of this study is to investigate ways to increase the number of knowledge creation projects that are commercialised, becoming the basis for innovation or further research by the industry.

Research Design and Methodology

The research is divided in two parts: In the first part, the study was focused on analysing the problem and defining its dimensions.

After defining the parameters of the problem, several alternative solutions were considered; one of these solutions was the introduction of post-project reviews. This possible solution was further investigated in the second part of the research;

1. Projects last for the time they are granted financing, in generally, one year. After that year, a new project can be started to continue the subject. In this way a set of successive projects can develop from idea to application. The commercialisation occurs in projects as well.

therefore, three research questions are formulated:

1. What is the right structure of the post-project review session?
2. What is the effect of the structure on the preparation and follow-up of the post-project review session?
3. How should the post-project reviews be introduced in the organisation?

Documentation and relevant theories in combination with knowledge obtained by means of interviews were the main sources of information necessary for answering these research questions. Subsequently, ten pilot-sessions were held; the sessions were analysed, and a survey was held among the participants of each session. The results of the analysis and the resulting knowledge became the basis for the final design of a methodology for the post-project review process. The drafted methodology was presented to the management who decided to implement it in all publicly funded knowledge creating projects.

Problem Analysis

The first part of the research focused on analysing and structuring of the motive of the research study: the lack of commercialisation of governmental funded projects.

Value Creation

The process of value creation consists according to Anderson and Narus (1999) of three phases: understanding value, creating value and delivering value. Applying this model to a research institute implies that knowledge is the value created. This knowledge is generated in knowledge-creating projects identified in the first stage of the process (understanding value). Delivering value implies the created knowledge-value finds its way to the market by means of commercialisation.

The reasons for failure to commercialise all knowledge created by the institute can be traced in each of the three phases of the value creation process (Anderson & Narus, 1999). Problems arising in the 'understanding value' phase can result in selecting projects of questionable technological interest and market potential. If problems arise in the 'creating value' phase, the projects can suffer from poor execution, and if problems arise in the 'delivering value' phase, the resulting innovations are not successfully brought to market.

Causes of Limited Success in Commercialisation

After analysing a number of the institute's projects in combination with employee interviews, the institute appears to be focused on the creating value phase. whereas not enough attention is paid to understanding and delivering value phases. This conclusion is based on the fact that many projects either lack a clear market focus (understanding value phase) and/or limited attention is paid to reach the customers interested in utilising the knowledge (delivering value phase). In other words, technological aspects are receiving much more attention in relation to market aspects. This can be attributed to various reasons; important ones are the internal technology-centred culture ignoring market needs and the lack of a consistent approach towards the value creation process from idea generation to market introduction (Kotler, 2003).

Increasing Market Focus and Awareness

Increasing the number of successfully commercialised knowledge-creating projects requires that the organisation changes its attitude as to the way it deals with the market and the customer needs as well as increasing peoples' awareness in the knowledge creation process. A way to stimulate such an attitude change is to encourage researchers to discuss not only the technical but also the commercial aspects of the application and pay special attention to the market potential of the project or set of projects. This requires thorough market orientation in the initial stages, resulting in a roadmap and thorough analysis of the results of each project, to map the next and future steps (Figure 2). With regard to the results analysis, a post-project review seems to be the proper way to achieve this. Currently, the institute does not apply post-project reviews. There are however reviews in between and the project managers are required to fill in a review form afterwards. The accent on these reviews is at the execution of the project (time, budget, risks).

Designing the Knowledge-Application Process Discussion

The design of the knowledge-application discussion process consists of a session and the preparation and follow-up of this session. Choices need to be made regarding the participants, the facilitator and the way to structure the session itself.

Post-Project Reviews in the Literature

In the literature, post-projects are mainly suggested as tools to facilitate and initiate organisational learning (von Zedtwitz, 2002). Busby (1999) concludes that post-project reviews are important learning tools, whose value is often underestimated. The post-project review is one of the most important, most structured and most broadly applicable ways to transfer knowledge (von Zedtwitz, 2002).

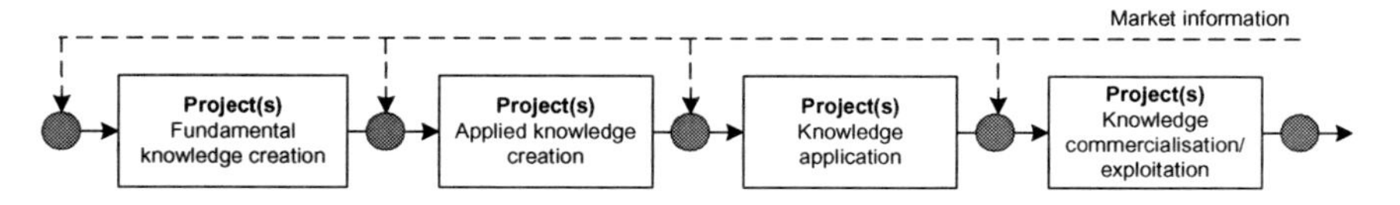

Figure 2: Innovation process with reflection moments.

Most organisations seem to lack a structural approach towards learning from past experience of projects. Even projects stopped prematurely are not always reviewed. A survey (von Zedtwitz, 2002) shows that 80% of projects are not reviewed afterwards, 20% are reviewed but without clear guidelines. Interim reviews are not uncommon, whereas many post-project reviews are only focused on technical aspects or skipped due to time and management restrictions (von Zedtwitz, 2002). The importance of post-project reviews and the fact that few organisations regularly carry them out are often underlined in the literature.

According to von Zedtwitz (2002), post-project reviews should focus on obtaining process information for future projects. The main goal is to initiate and facilitate the continuous learning on all levels within the organisation (focus on double-loop learning), which is crucial in R&D organisations. However, learning from reviews does not have to be restricted to the lifecycle of the project. von Zedtwitz (2002) gives an example of a post-project review in which new market potential is identified while technical knowledge is transferred to marketing employees. This is similar to the role the post-project review should be able to play for the institute.

Regarding the structure of the post-project review session, the approach chosen depends heavily on the existing company culture and underlying motive for conducting post-project reviews: different objectives and needs, different markets and industries, different cultural contexts, and different degrees of innovation all influence the way post-project reviews need to be conducted (von Zedtwitz, 2003).

Multiple Objectives

The main underlying motive to conduct the post-project market review in this case is the commercialisation of the projects' outputs; however, the institute does not have an explicit strategy for the way projects must be reviewed to contribute to the organisational learning. Therefore, the post-project review might have multiple objectives:

- Formal closing of the project by reviewing the course of the project for organisational learning.
- Discussing the application and commercialisation issues of the project as well as formulating the necessary course of action.

Of course, discussing the application and commercialisation of the projects' results is also something that can be done before and during the project. However, doing this in a structured way at the end of a project ensures that this step will not be omitted when the deliverables of the project are fixed. Besides, the probability that action decided and agreed during the session will be carried out increases when the project is over, since project activities will disrupt the agreements made.

The added value of this session based on the above premises, compared to the current situation, can be summarised in the following elements:

- The approach is compulsory and uniform for the entire institute.
- The approache encourages learning by reflection.
- It promotes identification of possibilities for application and commercialisation of the projects' results. These can be the input to the follow-up project.

Finally, the institute wants to introduce assessments for all projects. The post-project review seems a good occasion for this assessment. However, this can cause problems because the assessment can cause people feel bounded and are not honest and open about, for example, problems that appeared or about the potential of the projects' results.

People Involved

To reach the objectives of the sessions, the appropriate people need to be involved. In the case of learning by reflection, von Zedtwitz (2002) makes the distinction between three levels of learning: individual, team/group and organisational. A post-project review focuses on the learning between individual and team/group or/and the learning between team/group and the rest of the organisation. For the learning between individual and team/group, the entire project team needs to be present. For learning between team/group and the organisation, the acquired knowledge within the team needs to be transferred outside a team. This can be done in several ways. An effective way appears to be the presence of an outsider at a post-project review (Busby, 1999; von Zedtwitz, 2002). The outsider can be a project manager of similar project or someone from the Knowledge Management department. Knowledge Management can be an intermediary between the post-project reviews and (top) management.

The second goal, the commercialisation, requires some other participants, for example, customer manager, marketing manager or group manager. In the case of TNO Industrial Technology, the technology manager and sales manager should be involved. The technology manager has the overview over the (portfolio of) knowledge-creating projects (technology push); the sales manager is responsible for retaining the current customers and acquiring new ones (market pull). Together they can deliver a positive contribution to business development (Figure 3).

In conclusion, the following people should be participants of a post-project review session:

- The project team (including project manager).
- The technology manager of functional department.

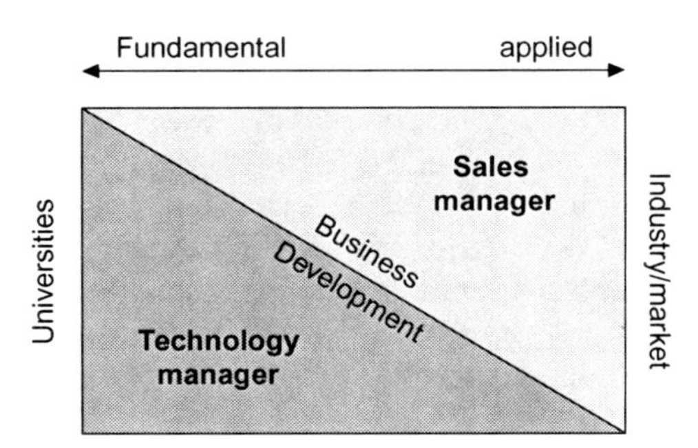

Figure 3: Role of sales manager and technology manager.

- The sales manager of a functional department.
- A representative of the Knowledge Management department.
- Others, for example, project managers of similar projects or the department manager.

This group of people is quite large and therefore needs to be reduced. As commercialisation is the most important goal, it is not required to involve the entire project team, only the key players are sufficient. Consequently, an additional meeting is required to facilitate the learning between individual and team/group.

The Facilitator

The course of the session depends largely on the facilitator. The facilitator can be the project manager or, for example, an outsider. Provided that they have the necessary experience and training, external facilitators have the advantage that they attend the meeting with an objective perspective. The external facilitator can be someone from the quality department (von Zedtwitz, 2002). In cases where this means someone from the knowledge management department, for example, will also be responsible for quality assurance. However, knowledge management is also responsible for the granting of government funding and, thus, the internal customer of knowledge-creating projects. Furthermore, the success of the session is also dependent on the motivation and support of all involved. This support is likely to be higher if the session is seen as a procedure required by their own department, rather than a staff department.

Within the department, the project manager can be the facilitator of the meeting but the disadvantage is that he is not objective at all. Most suitable of all is the technology manager. He has objectivity and is from the same department. Furthermore, as there is only one technology manager for every department and only seven functional departments, there are only seven technology managers. This means that they can be trained to facilitate future post-project reviews within their department.

Items on the Agenda of the Session

As the session has multiple independent goals, the sessions are split up according to these goals. It is easiest to start with the goal of organisational learning as this is looking back to the whole execution of the project. For this first part a number of questions (Table 1) are formulated, which are derived from the current project review form.

The second part of the session, the discussion about application and commercialisation of the projects' deliverables starts with the achieved deliverables of the project. This is already made clear during the first part of the session. The questions of Table 2 are formulated for this part of the session.

Table 1: Questions review.

	What Went Well or Badly?	Why?	How Could We Have Done It Differently?	What Can We Learn from This for Future Projects?
Did we reach are goals/ deliverables? Are our (internal) customers satisfied?				
How did the process go? (planning/actions/allocation of tasks/communication)				
How did deal with risks?				
Did we stay within the budget?				

Table 2: Questions discussion knowledge application.

What are our exact deliverables? What are the unique selling points (USPs) of these deliverables? What is the relationship with other similar projects and with the strategy of the institute and department? What is written down in the (existing) project plan about application and commercialisation of the deliverables? To what degree have activities for the application and commercialisation of the deliverables already started?	Preparation for discussion
How can we apply the knowledge created and then commercialise it? Are there other possibilities? (Feasibility study: market, size, competitors, prices, product) How can we do this? Which activities must be developed? (Marketing strategy) Who will do this? To do list with names and completion dates	Discussion

Finally, the session itself will be reviewed to continuously improve the session itself.

The assessment of the projects will be done after the session. This will cause as little as possible disturbance to free discussion.

Preparation and Follow-Up of the Session

To achieve the session objectives, it is important that the persons involves will prepare the session beforehand and that actions agreed will be followed up afterwards. The preparation requirements for all people involved are different. The project manager must have all the necessary information (process information, project report) available and distributed to the other participants. Next to that, the manager is the one initiating the session. The technology manager must ensure everything happens on time.

The follow-up is market oriented, and therefore, it is the responsibility of the sales manager. During the session, an action list with the steps that must be taken will be drafted: the sales manager is responsible for these steps.

Pilot Sessions

Before permanent implementation of the pilot-sessions, 10 pilot-sessions were organised. The pilot-sessions should give more insight and understanding of the process and, based on the results, adjustments could be made. It was also an opportunity to confront the employees with the concept before becoming a standard organisational process. Furthermore, based on the results of the pilot-sessions, the management team can decide whether to continue or not.

Execution

To make possible to carry out the pilot sessions in short term, some changes were to the previously described design. Most important is that knowledge management initiated and facilitated the sessions, because the technology managers are not trained yet in chairing the sessions. Knowledge Management cooperated in developing the sessions and had full knowledge of details and reasoning.

The projects that were selected for the pilot sessions were from the different functional departments. Nine of ten projects were fully funded by the government grants, one project was funded for 25% by a commercial organisation. All projects were completed; it was known that some projects would be followed-up by a subsequent project.

For all projects, a meeting was organised with the participation of the project manager, the technology manager, the sales manager and a representative of knowledge management. The project manager was encouraged to invite key players from the project team as well. In one department, the department manager carries out the role of sales manager and technology manager. Therefore, only three people were present at those sessions.

To evaluate the pilot-sessions, these were observed and the participants were asked for their opinion. The focus of both the observations and the questionnaire was on the extent to which the objectives of the session were met and if not what were the possible reasons for that.

After two sessions, it was already clear that a single session for both the project review and the discussion of knowledge application were ineffective. The main reason for this was a defensive attitude of the project manager after the first part of the session, reflection on organisational learning. The defensive attitude seems to stem from the project assessment part and the facilitation by the board member responsible for technology (representative of the Knowledge Management department). The fact that it was the first confrontation with the post-project review might have had some influence as well.

As such a defensive attitude is not desirable, from the third pilot-session on; the sessions are strongly focused on discussing the knowledge application and less focused on a project review. The project review was reduced to one question at the end of the meeting: Hence, the session is called 'knowledge application discussion'.

Results

As mentioned, the results were determined by observation and feedback from the participants.

Observation indicated a lot of variation between different sessions. Various aspects caused the differences. One of these aspects was the nature of the project; some projects are more fundamental, others more applied (Table 3). This resulted in different discussions during the sessions. Discussions during sessions of more

Table 3: Results observations pilot-sessions: distinction caused by the nature of a project.

	Application and Commercialisation Not Clear	**Application and Commercialisation Fairly Clear**
Succession	Project with 100% governmental funding	Sometimes partly governmental funding
Discussion about	Applications of result of both projects (current and next), roadmap	Next steps (to do's) for commercialisation; alternative options
Added value of the session	Stimulation to think about it	Moment for reflection of all possibilities, participation of different people
Concrete to do points (for example)	Clarify roadmap for next project in a next meeting	Approaching specified organisations by sales managers

fundamental projects were focused on the possibilities for applying the results. These projects in general already had a follow-up project and although funding for the follow-up project was granted, the application and commercialisation of the projects' outcomes was not considered. This resulted during the sessions in a discussion for direction of the follow-up project. The session added value to the process because the question about the application and commercialisation of the results were brought up and the researchers were forced to think about it. An example of a concrete deliverable of the sessions was an appointment for further development of the roadmap.

Discussions during sessions of more applied projects involved the commercialisation of the created knowledge, the application of the knowledge was already known. The added value of the session was originating from the new insights of the 'outsiders' and by the stimulation to explore all commercial possibilities. Concrete deliverables of these sessions was, for example, a to-do-list with actions like the approach of specified organisations by the sales manager.

The atmosphere was another aspect that made a difference between the sessions. A good atmosphere proved vital to reach the goals of the sessions, however, during two sessions participants did not feel very motivated. Both sessions had no designated deliverables, all others had.

The results from the questionnaire were positive. Ninety-four percent said that the sessions were useful, and 71% said that they thought that the sessions in general would improve the number of projects that will be commercialised (Table 4).

Final Design Post-Project Market Review

After the pilot-sessions, the management was advised to continue the sessions (knowledge application discussion), and they agreed to this for all projects with 100% governmental funding. The sessions would be held in the same form as the pilot-sessions with some minor improvements. Furthermore, the sessions would eventually be facilitated by the technology managers of the departments and be initiated by the project manager. The introduction of the session as a standard procedure would be done gradually. One important point of attention is the

Table 4: Summary results questionnaire pilot-sessions.

Question/Thesis	**Agree (%)**	**Disagree (%)**
Do you think the session was useful?	94%	6%
'Discussions about knowledge application will contribute to a conscious evaluation of the innovation process'	93%	7%
'Discussions about knowledge application will improve the number of project results that are commercialised afterwards'	71%	29%
'Project reviews will improve organisational learning'	80%	20%

motivation of the participants in the sessions. All technology managers should be convinced of the usefulness, and this must be communicated thoroughly to all other participants.

Conclusions and Discussion

In this case study, the concept of post-project review is used as a tool to stimulate the commercialisation of new technologies. The essence of the knowledge-application discussion is to bring multi-functional and multilevel participants together at the end of a knowledge-creating project to discuss the application and commercialisation of the project results. For projects followed up by a new (wholly or partly governmental funded) project, the knowledge application discussion resulted in a framework for the direction of the next project; hence, a post-project review session — before the follow-up project begins — can become a stimulus to the innovation process. In this sense the session can be seen as a moment of reflection on the direction taken with regard to future market opportunities. During the innovation process, the discussion will develop from a discussion about the application itself to the commercialisation of it.

By involving the technology managers and knowledge management, the sessions are also becoming tools to relate projects or innovation processes (groups of projects) to each other and widen their scope: the knowledge-creating projects are this way not limited to a single discipline in the chain from fundamental to applied knowledge, but can expand across disciplines and research areas.

Next to the session, it is also necessary to reflect on the value of the innovation at the beginning of the innovation process; a suitable moment for reflection is the submission of the request for government funding; the request must also be based on future market opportunities. In this case this means changing the current attitude towards the granting of procedures and criteria so that the organisation is able follow the line of increasing the chances of focussing on commercially interesting projects. This will require, among other things, a more extensive market exploration.

For further research, the following questions will be interesting:

- How common are knowledge-application discussion sessions in organisations and how are they carried out with regard to objectives, participants, facilitation and items on the agenda?
- During the pilot sessions, the combination of reviewing and discussing the application did not seem to work; should these be two, separated discussion items?
- Furthermore, what other tools are used to tackle the lack of commercialisation of governmental funded projects?

References

Anderson, J. C., & Narus, J. C. (1999). *Business market management, understanding, creating and delivering value*. Upper Saddle River, NJ: Prentice-Hall.

Busby, J. S. (1999). An assessment of post-project reviews. *Project Management Journal, 30*(3), 23–29.
Kotler, P. J. (2003). *Marketing management* (11th ed.). Upper Saddle River, NJ: Prentice-Hall.
von Zedtwitz, M. (2002). Organizational learning through post-project reviews in R&D. *R&D Magazine, 32*(3), 255–268.
von Zedtwitz, M. (2003). Post-project reviews in R&D. *Research-Technology Management* (September–October), 43–49. Industrial Research Institute, Inc.

Chapter 8

Using Knowledge Management to Gain Competitive Advantage in the Textile and Apparel Value Chain: A Comparison of Small and Large Firms

Paula Danskin Englis, Basil G. Englis,
Michael R. Solomon and Laura Valentine

Knowledge and Knowledge Management

Knowledge theories have developed over the past 30 years (Polanyi, 1966). However, it is only recently that knowledge has become regarded valuable asset in corporate boardrooms. Knowledge acquisition has become a critical resource for creating and sustaining competitive advantage as the competitive environment continues to intensify (Hitt, Ireland, & Lee, 2000). As with other corporate assets, the processes surrounding the creation and transfer of knowledge must be managed with significant insight to derive the most value from knowledge investments (Bhagat, Kedia, Harveston, & Triandis, 2002; Conner & Prahalad, 1996; Davenport & Prusak, 1998; Edvinsson & Malone, 1997; Stewart, 1997). The purpose of this chapter is to examine the significance of managing knowledge both within firm (internal knowledge) and across the value chain (external knowledge) for small and large firms. First, we review the literature on knowledge management systems and propose some hypotheses for internal and external knowledge management. Next, we present the data and follow this with the results. Discussion of the results follows, and the chapter closes with a number of managerial implications, limitations, and suggestions for future research.

The quest to innovate through research and development is essential for firms to remain ahead of competitors. Indeed, many firms view the acquisition of new

New Technology Based Firms in the New Millennium, Volume VII
Edited by R. Oakey, A. Groen, G. Cook and P. van der Sijde

knowledge as a way to gain and maintain competitive advantage (Danskin, Englis, Solomon, Goldsmith, & Davey, 2005). However, few firms fully realize the benefits from high value knowledge. Knowledge that is isolated in one department or in a specific segment of the value chain is not utilized to its full extent. New knowledge should be harnessed and managed through internal knowledge management systems that create learning opportunities for other departments or product areas within the firm. Internal knowledge management systems may provide platforms for the further development of knowledge transfer to external partners. By implementing internal and external knowledge management systems, firms can experience a greater competitive advantage and sustain success over a longer period of time.

Types of Knowledge Management Systems

There are two general types of knowledge management systems that firms use to provide a basis for renewing competitive advantage. First, passive knowledge management systems (such as the EDI system used by Wal Mart) are distinguished by their orientation to the "present" and tend to be used with channel members such as suppliers to more closely schedule component deliveries, reduce cycle time, cut inventories, and decrease the overall costs of production based on current behavior of buyers and sellers. Second, in contrast, active knowledge management systems have a "future orientation" and tend to be used by individuals to add value to the product as it passes through value chain. Active knowledge management systems not only reap the benefits of reduced costs and cycle time but also develop valuable knowledge that anticipates of future buyer/seller behavior (e.g., market backed research and development). Proactive knowledge management systems do not only enhance efficiency through time and cost savings; they also provide a way to link and leverage the "voice of the consumer" to all stages of product development, production, and distribution through the value chain. While anecdotal evidence suggests that some firms are building knowledge management systems that include both active and passive orientations to provide feedback loops throughout the value chain, there is no empirical research relating these developments to strategy, the value chain position, and firm performance.

Knowledge Management Systems — Internal Processes

The effectiveness of building knowledge within the firm depends on the firm's ability to monitor and absorb newly acquired knowledge from many sources and integrate this knowledge into its existing knowledge base (Hamel, 1991; Hansen, Nohria, & Tierney, 1999). Internal knowledge management systems can also be thought of as organizational memory. Establishing organizational memory through knowledge management systems is an essential task to be completed before firms venture into knowledge sharing with value chain partners. Before developing knowledge management systems, businesses need to understand the process of organizational memory. As shown in Figure 1, this process is divided into four separate parts comprising acquisition, retention, maintenance, and retrieval (Stein, 1995).

As mentioned above, part of internal knowledge management involves organization memory. Acquisition and retention play key roles in this process. Acquisition involves both internal and external research and development. Innovation or new knowledge facilitates value-added product development that leads to an increase in competitiveness. Retention of organizational knowledge typically involves developing processes, procedures, and systems. In this way, retention can be thought of as a codification process designed to create organizational memory. The network in process of some firms involves the use of databases that record knowledge for future use, whereas, other firms may have an organizational culture in which knowledge is shared by informal mechanisms such as talking at the water cooler or at the coffee pot. While informal networks retain knowledge at a higher rate than distributed information system, such knowledge is not easily maintained for future use. Retention is facilitated by three mechanisms. Theses mechanisms are schemas, scripts, and systems. The importance of harnessing internal knowledge cannot be underestimated. Small firms may lack the time, money, or other resources needed to develop a knowledge retention system. As firms grow larger, they generally build internal systems and structures to manage the flow of information across the firm. Therefore, we expect that smaller firms will have fewer resources to develop and establish internal knowledge management systems, particularly those that facilitate organizational memory. The above discussion suggests a numbers of hypotheses to be tested.

Hypothesis 1. Large firms will have a more developed organizational memory than smaller firms.

A second aspect of the internal knowledge management process involves the role of maintenance and retrieval or organizational memory. Indeed, the maintenance of knowledge is often overlooked when discussing organization memory. However, if knowledge is not properly maintained, information can become misconstrued or lost all together. When knowledge is stored in databases, maintenance is simple, although when information is stored within informal networks using individual minds, the maintenance becomes complicated. This is especially true for employee turnover, when valuable knowledge leaves with the former employee and is not transferred

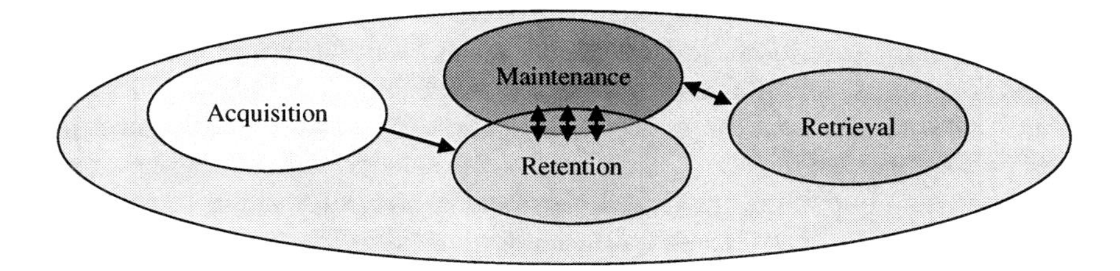

Figure 1: Internal organizational memory.

back to new employees. Of particular importance is the role of experts. When experts leave the firm, they take their knowledge and their informal knowledge network with them, which can be damaging to firm competitiveness (Prahalad & Hamel, 1990). The retrieval of knowledge is one of the most important aspects of organizational memory. Managers should develop support mechanism, motivation, and rewards for knowledge sharing and retrieval to be successful. Individuals must be motivated to retrieve and communicate information. Ernst & Young, for example, evaluates and rewards its employees based on their contribution to the knowledge of the firm (Hansen et al., 1999). A major problem within many organizations is the fact that employees view knowledge as a method of securing their jobs and are consequently reluctant to share their knowledge. The retrieval of internal knowledge across the firm can facilitate the discovery and exploitation of opportunities. Internal knowledge may lead to a technological breakthrough that represents an opportunity despite its market applicability not being initially apparent (Abernathy & Utterback, 1978). This knowledge can also enhance a firm's ability to effectively exploit an opportunity by, for example, determining a product's optimal design to optimize functionality, cost, and reliability (Rosenberg, 1994) and ultimately the economic impact of exploiting the opportunity (McEvily & Chakravarthy, 2002). Therefore, the ability to retrieve internal knowledge provides a firm with the ability to rapidly exploit opportunities or to be able to respond quickly when competitors make advances (Cohen & Levinthal, 1990).

From the above, we expect that larger firms with more resources will focus on internal knowledge systems and structures more than smaller firms. More developed internal knowledge management systems will enable people across the firm to more fully access internal knowledge for market applicability and new opportunities. Thus, the following hypothesis is offered.

Hypothesis 2. Organizational memory will be more dispersed in large firms than small firms.

Knowledge Management Systems — External Processes

External knowledge management systems are often comprised of Internet-based systems that link members of a value chain. On a functional level, external knowledge management systems are transparent and allow every member of the value chain to "see" the operations of every other member through production schedules, shipping schedules, ordering schedules, and inventory levels. At a strategic level, knowledge management systems, when shared across the value chain, bring the "voice of the consumer" very clearly into the process. This allows the entire value chain to view changing customer preferences. The early knowledge of changing consumer preferences creates opportunities for all members of the value chain to react, almost immediately, thus reducing the cycle time of product development and change.

External knowledge management has received increasing attention from the academic community (Andersen & Christensen, 2000; Bessant, 2004; Dyer & Singh,

1998; Dyer & Nobcoka, 2000; Hakansson, Havila, & Pedersen, 1999; Sako, 1999; Hult, Ketchen, & Slater, 2004; Wagner & Bukó, 2005). Most of this research has been conceptual to date. For instance, Dyer and Singh (1998) suggest that value chain relationships are significantly affected by the learning and sharing of knowledge. Exceptions include case studies by Andersen and Christensen (2000) and Hakansson et al. (1999). These case studies show that firms tend to learn and share more knowledge when they are embedded in a network — such as a supply chain. Larger firms may have more structured systems that emphasize learning used to tap into their knowledge networks. These external knowledge management systems can lower costs tremendously by increasing communication and eliminating steps in the manufacturing process that are unnecessary. For instance, the Toyota corporation uses this type of system to emphasize knowledge sharing with its supplier networks (i.e., Kogut, 2000). Firms can gain significant benefits from integrating knowledge from external sources outside the firm (Dyer & Nobcoka, 2000; Kogut, 2000; Mohr & Sengupta, 2002). Value chain partners can also experience rapid learning by accessing onto another's learning curve for particular processes or procedures such as Six Sigma Continuous Improvement. Knowledge sharing leads to increased quality and heightened customer perceptions of brand platforms. Such knowledge stores can be accessed through interorganizational relationships with customers, suppliers, and other bodies outside the company (Dyer & Singh, 1998; Madhok & Tallman, 1998). Schroeder, Bates, and Junttila (2002) have found that that external learning and knowledge transfer among the firms and their suppliers and customers is the strongest contributor to manufacturing performance in their empirical study of 164 manufacturing plants. Learning and sharing knowledge with suppliers play an important role in interfirm buyer–supplier relationships (Dyer & Singh, 1998; Sobrero & Roberts, 2002). Suppliers may possess resources that complement the firm's knowledge base, which may generate positive externalities and allow the firm to capture "spill overs" from its suppliers (Lorenzoni & Lipparini, 1999). Based on our review of the literature, we expect that the ability to establish an external knowledge management system to learn from the others in the value chain is likely to result in sustained competitive advantages for the firm. Based on our review of the literature, larger firms are more likely than smaller firms to focus on learning from value chain members.

Hypothesis 3. Larger firms are likely to have external knowledge management systems that emphasize learning more than smaller firms.

We expect that larger firms will also focus on developing external knowledge management systems that foster innovation with value chain partners more than smaller firms. Larger firms will be more likely to standardize practices, processes, and platforms among value chain partners. This drive for uniformity across the value chain increases knowledge sharing, cooperative developments, and the utilization of information captured from supply chain systems. The more developed the external knowledge management systems becomes the more likely the firm will learn from partners' knowledge concerning market applicability and new opportunities. We expect that smaller firms will also focus on developing external knowledge management systems that foster entrepreneurship activities with value chain partners more than larger firms. Smaller firms are more likely to be entrepreneurially focused than larger firms and more able to take advantage of entrepreneurial opportunities. They are likely to adopt the latest supply chain technologies and may engage in higher risk projects. We expect larger firms will focus more on innovation and smaller firms will focus more on entrepreneurship.

Hypothesis 4. Larger firms are likely to have external knowledge management systems that emphasize innovation more than smaller firms.

Hypothesis 5. Smaller firms are likely to have external knowledge management systems that emphasize entrepreneurship more than larger firms.

Methodology

The goal of this research was to develop a descriptive framework and explore possible relationships among variables (Campbell & Stanley, 1963). The design selected was a non-experimental, static group comparison survey that is suitable for exploratory investigations where a phenomenon is described (Denzin, 1978).

Sample

To allow for maximum generalizability, a national sample of US firms participating in the apparel and textile industries was used. We chose this industry because it has come under severe international competition in the past decade and many low cost participants have moved operations overseas. We expected that firms in this industry would be forced to compete on other factors such as knowledge management. A US national sample reduces any bias that misuse occurs due to economic variations in certain areas of the country. The sample was drawn from a database maintained by InfoUSA, an information services company located in Boston, MA. The database contained archival information on all firms in the sample and was used to compare the groups across broad categories (total sales, year the firm was founded, and number of employees) to test for non-response bias. The firms in the sample competed in many segments along the value chain of the US textile and apparel industries.

Survey

The major method of data gathering was an online survey. The survey was developed inductively using existing scales that were slightly modified for the specific purpose of this study. Pre-testing was used to check the questionnaire for comprehension and content validity. The instrument was evaluated by a group of academic experts and a practitioner from the National Council of Textile Organizations. This group reviewed and commented on issues such as clarity, order of questions, comprehensiveness and parsimony, and overall presentation of questionnaire. Efforts to increase the response rate were taken including offering to send respondents an executive summary of the results (Hinrichs, 1975), and the survey was emailed during a non-holiday period. The survey was also reviewed and approved by the Institutional Review Board at Berry College.

The survey was sent to members of the top management team of the firm since previous studies have found that top executives have relevant information about the strategy of the firm (Hambrick & Mason, 1984) and value chain management (Kobrin, 2000). Of the 310 people who work in textile and apparel industries to whom we sent the survey, 32 completed it resulting in a response rate of 10.32%. This sample was used to test the internal consistency of the measures. We are currently collecting more data using a larger sample of 2535 managers in the textile and apparel industry value chain.

Measures

Internal Knowledge Management

Research on internal has focused on two main areas: organizational memory level and organization memory dispersion. Before answering questions on internal knowledge management, respondents were first asked to think about a specific new project that they were familiar with that recently occurred within their firm. The respondents were asked to keep this project in mind when answering questions about internal knowledge. Organizational memory (ORGMEM) is defined as the amount of stored information or experience an organization has about a particular phenomenon (Moorman & Miner, 1997). It was measured by asking respondents to answer four questions on a seven-point Likert scale, where 7 = strongly agree and 1 = strongly disagree. Respondents were asked, "Prior to the project, compared to other firms in our industry, my division had "a great deal of knowledge about the category," "a great deal of experience in the category," "a great deal of familiarity with the category," and "invested a great deal of R&D in this category." The responses to these questions were subjected to exploratory factor analysis using principal component analysis and were tested for reliability through Cronbach's alpha (Nunnally, 1978). All items loaded on the same factor (Eigenvalue = 2.82) and the reliability was consistent with previous studies (Cronbach alpha = 0.85, $N = 32$).

The second component of internal knowledge management is organizational memory dispersion (MEMDIS). Memory dispersion refers to the degree to which organizational memory is shared throughout the relevant organizational memory unit. If memory is widely shared, memory dispersion is high. If memory is not widely shared, memory dispersion is low. Respondents were asked to rate on a seven-point scale where "7 = high" and "1 = low," the degree of consensus among the people working on the project for the following new product areas consisting of product design, brand name, packaging, promotional content, and product quality level. The responses to these questions were subjected to exploratory factor analysis using principal component analysis and were tested for reliability through Cronbach's alpha (Nunnally, 1978). All items loaded on the same factor (Eigenvalue = 3.39) and the reliability was acceptable (Cronbach alpha = 0.88, $N = 32$).

External Knowledge Management

Three constructs pertaining to external knowledge management were adapted from Hult, Ketchen, & Nichols (2002). Supply chain innovativeness (SCINN) is continuous improvement through creativity and ingenuity (Hult et al., 2002). Generally, firms possessing innovativeness will strive to not only meet customer's current needs but also anticipate future needs. This construct was assessed on a seven-point Likert scale where "1 = strongly disagree" and "7 = strongly agree." Respondents were asked to click on the response that best indicates the extent of your agreement with each statement below: "Technical Innovation, based on research results, is readily accepted in the supply chain," "We actively seek innovative supply chain ideas," "Innovation is readily accepted in the supply chain process," "People are not penalized for new supply chain ideas that do not work," and "Innovation in our supply chain is encouraged." The responses to these questions were subjected to exploratory factor analysis using principal component analysis and were tested for reliability through Cronbach's alpha (Nunnally, 1978). All items loaded on the same factor (Eigenvalue = 3.44) and the reliability was acceptable (Cronbach alpha = 0.88, $N = 32$).

The second external knowledge management is supply chain learning (SCLEARN). This is the generation of new insights that have the potential to change behavior gained from other value chain members (Huber, 1991; Hult et al., 2002). This construct was assessed on a seven-point Likert scale where "1 = strongly disagree" and "7 = strongly agree." Respondents were asked to click on the response that best indicates the extent of your agreement four items were listed, "The sense around here is that employee learning is an investment, not an expense in the supply chain," "The basic values of this supply chain process include learning as a key to improvement," "Once we quit learning in the supply chain we endanger our future," and "We agree that our ability to learn is the key to improvement in the supply chain process." The responses to these questions were subjected to exploratory factor analysis using principal component analysis and were tested for reliability through Cronbach's alpha (Nunnally, 1978). All items loaded on the same factor (Eigenvalue = 3.20) and the reliability was acceptable (Cronbach alpha = 0.91, $N = 32$).

The third and final component of external knowledge management is supply chain entrepreneurship (SCENT). Entrepreneurship in the context of the supply chain is defined as pursuit of new market opportunities and the renewal of existing areas of an organization's operations (Hult et al., 2002). This construct was assessed on a seven-point Likert scale where "1 = strongly disagree" and "7 = strongly agree." Respondents were asked to click on the response that best indicates the extent of your agreement. There were five items: "We believe that wide-ranging acts are necessary to achieve our objectives in the value chain," "We initiate actions to which other organizations respond," "We are fast to introduce new administrative techniques and operating technologies in the supply chain," "We have a strong proclivity for high risk projects in the supply chain," and "We are bold in our efforts to maximize the probability of exploiting opportunities in the supply chain." The responses to these questions were subjected to exploratory factor analysis using principal component analysis and were tested for reliability through Cronbach's alpha (Nunnally, 1978). All items loaded on the same factor (Eigenvalue = 2.78) and the reliability was acceptable (Cronbach alpha = 0.79, $N = 32$).

Data Analysis and Results

The first set of analysis involved examining a listwise correlation (Table 1) among all variables for the sample ($N = 32$). In this research, correlation analysis showed several of the correlations were significant indicating that additional analyses were warranted. To test the hypotheses, a second set of analyses (*t*-tests) examined the mean differences for the involved variables between small and large firms. The sample was broken into two groups based on the average sales of the firms ($500,000). There were 18 small firms and 14 large firms.

The first set of hypotheses, Hypothesis 1 and Hypothesis 2, predicted differences between internal knowledge management practices of small and large firms competing in the textile and apparel value chain (organizational memory, organizational knowledge dispersion). Overall, the results provide support for these hypotheses regarding differences between small and large firms, which have that larger firms would have more developed organizational memory than smaller firms (Hypothesis 1) and that organizational memory would be more dispersed in larger firms than smaller firms. The first hypothesis was supported. Our results show that organizational member is significantly higher ($p = 0.09$) in larger firms (mean 13.57) than smaller

Table 1: Correlations.

	ORGMEM	MEMDIS	SCINN	SCLEARN	SCENT
ORGMEM					
MEMDIS	.400*				
SCINN	.068	.207			
SCLEARN	−.044	.319	.617**		
SCENT	.331	.035	.147	.195	

*Correlation is significant at the 0.05 level (2-tailed). **Correlation is significant at the 0.01 level (2-tailed).

firms (mean 10.05). On the contrary, the second hypothesis was not supported. Although larger firms did have higher levels of organizational memory dispersion (mean 14.21), this was not significantly different to that of small firms (mean 12.33).

The second set of hypotheses (Hypothesis 3, Hypothesis 4, and Hypothesis 5) predicted differences between external knowledge management practices of small and large firms competing (supply chain innovation, supply chain learning, and supply chain entrepreneurship). The results were mixed. In terms of supply chain learning, the results showed that larger firms did have higher levels of learning (mean 10.35 vs. 8.61). However, these differences were not significant. Thus, Hypothesis 3 was not supported. For supply chain innovation, we proposed that larger firms would emphasize innovation more than small firms (Hypothesis 4). This hypothesis was supported. Our results show that supply chain innovation was significantly higher ($p = 0.05$) in larger firms (mean 18.71) than for smaller firms (mean 14.33). The final hypothesis was not supported. We proposed that smaller firms would have higher levels of supply chain entrepreneurship. Results show that small firms' level of entrepreneurship (mean 16.16) was not significantly different than large firms (mean 17.71).

Discussion and Conclusions

The difficulties of managing knowledge are faced by firms of all sizes. The purpose of this research was to examine knowledge management systems within the firm through organizational memory and outside the firm through innovation, learning, and entrepreneurship across the value chain. Specifically, we proposed that small firms manage knowledge differently than large firms.

Our results show that large firms differ significantly from small firms in how they manage knowledge both internally and externally. Larger firms have significantly more developed organizational memory systems. However, small firms are just as good as their larger counterparts at dispersing organizational memory or sharing information with employees across the firm. Survey results indicate that smaller firms may not require formal knowledge structures to preserve knowledge. Small size may facilitate informal mechanisms such as meetings around the water cooler or around the coffee pot to share internal knowledge. Small firms also do not have such distinct hierarchal structures, or the fierce departmental rivalries, seen within large organizations, that thwart internal knowledge management.

In terms of external knowledge management, large firms emphasize supply chain innovation more than smaller firms. This may be due to increasing pressures in the textile and apparel value chain to cut cycle time. Larger firms generally coordinate longer portions of the value chain than smaller firms, thus facing increased pressure to innovate and decrease cycle times among several firms. Large firms also tend to have more expertise specific to supply chains at their disposal and have significantly more capital to fund supply chain projects.

The goal of our research was to understand more about knowledge management and how the process of acquisition, retention, maintenance, and retrieval of knowledge, both within the firm, by improving organizational memory, and across the value chain through knowledge management systems, may help firms gain competitive advantage. This research will also help both small and large firms to examine and develop their knowledge management systems internally and externally. Internal systems create and sustain organizational memory. Organizational knowledge such as routines and processes are more easily stored whereas tacit knowledge of key individuals is much more difficult to codify. Organizational memory creates opportunities to minimize knowledge isolation in functional departments and creates a greater base from which tacit learning can be derived. Firms with robust organizational memories are less damaged when key personnel leave. External knowledge management systems bring value chain members closer together and add value to products (e.g., increased quality, customer perceptions of brand platforms) throughout the value chain. Opportunities for innovation increase as partners discover new possibilities or combinations of knowledge put into the value chain processes. These opportunities may decrease the costs of products or create innovative new applications for mature products. The overall impact of knowledge management systems engaged across the value chain is to render superior image products to low cost substitutes in the marketplace and create sustainable competitive advantage for all partners.

Managerial Implications

From a managerial perspective, this study has several important implications. First, managers need to create and manage both internal and external knowledge management systems whether they are active or passive in nature. Internal systems are important as means of codifing and creating organizational memory. They also facilitate the dispersion of knowledge across the firm giving employees a fuller picture of the firm's knowledge base. While larger firms have more resources to create and store internal knowledge, small and large firms were equally good at dispersing knowledge across the firm. Managers should also manage knowledge sharing in their supply chain (i.e., customers, suppliers and manufacturers, mills) by committing sufficient resources to setting up, maintaining, and monitoring a knowledge sharing network. While managers of larger firms may have more resources at their disposal to create these networks, managers of small firms can nonetheless benefit from supply chain networks. Our research shows that small firms are equally as capable of innovating through the value chain as their larger counterparts and have similar levels of entrepreneurship gained through value chain interaction.

Limitations and Suggestions for Future Research

The results presented here are subject to some limitations. First and perhaps most important, the results were based on a very small sample from a single industry.

However, since this data collection is not yet complete, we hope to confirm and extend these results during future analysis. A second limitation is the use of a single respondent per firm, which did not allow us to ascertain whether any of the respondent firms had value chain partners. A longitudinal study of knowledge sharing networks would be an excellent addition to this body of literature. We also looked only at differences in knowledge management systems in large and small firms and did not link this information to firm performance. Since we propose that knowledge is strategically important and can be a source of competitive advantage, we recommend that further research be conducted to tie knowledge management systems to multiple forms of performance including both financial and cycle time performance implications.

Other areas that offer some interest include examining the role of the absorptive capacity and firm culture (Cohen & Levinthal, 1990; Levinson & Asahi, 1995). It may also be interesting to investigate the use of knowledge management tools, shared communication vehicles, and the facilitation of information technology since they may augment our understanding of internal knowledge management and external knowledge sharing.

This work was supported by the Richard Edgerton Endowment and the National Textile Center.

References

Abernathy, W. J., & Utterback, J. (1978). Patterns of industrial innovation. *Technological Review*, *80*(7), 40–47.

Andersen, P. H., & Christensen, P. R. (2000). Inter-partner learning in global supply chains: Lessons from NOVO Nordisk. *European Journal of Purchasing & Supply Management*, *6*(2), 105–116.

Bessant, J. (2004). Supply chain learning. In: R. Westbrook & S. New (Eds), *Understanding supply chains: Concepts, critiques, futures* (pp. 165–190). Oxford, England: Oxford University Press.

Bhagat, R. S., Kedia, B. L., Harveston, P., & Triandis, H. C. (2002). Cultural variations in the cross-border transfer of organizational knowledge: An integrative framework. *Academy of Management Review*, *27*(2), 1–18.

Campbell, D. T., & Stanley, J. C. (1963). Experimental and quasi-experimental designs for research on teaching. In: N. L. Gage (Ed.), *Handbook of research on teaching*. Chicago: Rand McNally.

Cohen, W. M., & Levinthal, D. A. (1990). Absorptive capacity: A new perspective on learning and innovation. *Administrative Science Quarterly*, *35*(1), 128–152.

Conner, K. R., & Prahalad, C. K. (1996). A resource based theory of the firm: Knowledge versus opportunism. *Organization Science*, *7*, 477–501.

Danskin, P., Englis, B. E., Solomon, M. R., Goldsmith, M., & Davey, J. (2005). Knowledge management, the value chain and competitive advantage: Lessons from the textile industry. *Journal of Knowledge Management*, *9*(2), 91–102.

Davenport, T. H., & Prusak, L. (1998). *Working knowledge*. Boston: Harvard Business School Press.

Denzin, K. (1978). *The research act*. New York: McGraw-Hill.
Dyer, J. H., & Nobcoka, K. (2000). Creating and managing a high performance knowledge-sharing network: The Toyota case. *Strategic Management Journal, 21*(3), 345–367.
Dyer, J. H., & Singh, H. (1998). The relational view: Cooperative strategy and sources of inter-organizational competitive advantage. *Academy of Management Review, 23*(4), 660–679.
Edvinsson, L., & Malone, M. S. (1997). *Intellectual capital: Realizing your company's true value by finding its hidden brainpower*. New York: Harper Business.
Hakansson, H., Havila, V., & Pedersen, A.-C. (1999). Learning in networks. *Industrial Marketing Management, 28*(5), 443–452.
Hambrick, D. C., & Mason, P. A. (1984). Upper echelons: The organization as a reflection of its top managers. *Academy of Management Review, 9*, 193–206.
Hamel, G. (1991). Competition for competence and interpartner learning within international strategic alliances. *Strategic Management Journal, 12*, 83–103.
Hansen, M. T., Nohria, N., & Tierney, T. (1999). What's your strategy for managing knowledge? *Harvard Business Review, 77*(2), 106–116.
Hinrichs, J. (1975). Factors related to survey response rates. *Journal of Applied Psychology, 60*, 249–251.
Hitt, M. A., Ireland, R. D., & Lee, H. (2000). Technological learning, knowledge management, firm growth and performance: An introductory essay. *Journal of Engineering and Technology Management, 17*, 231–246.
Huber, G. (1991). Organizational learning: The contributing processes and literature. *Organization Science, 2*, 88–115.
Hult, G. T. M., Ketchen, D. J., & Nichols, E. L. (2002). An examination of cultural competitiveness and order fulfilment cycle time with supply chains. *Academy of Management Journal, 45*(3), 557–586.
Hult, G. T. M., Ketchen, D. J., & Slater, S. F. (2004). Information processing, knowledge development, and strategic supply chain performance. *Academy of Management Journal, 47*(2), 241–253.
Kobrin, S. J. (2000). Development after industrialization: Poor countries in an electronically integrated global economy. In: N. Hood & S. Young (Eds), *The globalization of multinational enterprise activity and economic development*. New York: St Martin's Press, Inc.
Kogut, B. (2000). The network as knowledge: Generative rules and the emergence of structure. *Strategic Management Journal, 21*(3), 405–425.
Levinson, N. S., & Asahi, M. (1995). Cross-national alliances and interorganizational learning. *Organizational Dynamics, 24*(2), 50–63.
Lorenzoni, G., & Lipparini, A. (1999). The leveraging of interfirm relationships as a distinctive organizational capability: A longitudinal study. *Strategic Management Journal, 20*(4), 317–338.
Madhok, A., & Tallman, S. B. (1998). Resources, transactions and rents: Managing value through interfirm collaborative relationships. *Organization Science, 9*(3), 326–339.
McEvily, S. K., & Chakravarthy, B. (2002). The persistence of knowledge-based advantage: An empirical test for product performance and technological knowledge. *Strategic Management Journal, 23*(4), 285–305.
Mohr, J. J., & Sengupta, S. (2002). Managing the paradox of interfirm learning: The role of governance mechanisms. *Journal of Business and Industrial Marketing, 17*(4), 282–301.
Moorman, C., & Miner, A. S. (1997). The impact of organizational memory on new product performance and creativity. *Journal of Marketing Research, 34*(February), 91–106.
Nunnally, J. C. (1978). *Psychometric theory* (2nd ed.). New York: McGraw-Hill.

Polanyi, M. (1966). *The tacit dimension*. London, UK: Routledge and Kegan Paul.
Prahalad, C. K., & Hamel, G. (1990). The core competence of the corporation. *Harvard Business Review* (May–June), 79–91.
Rosenberg, N. (1994). *Exploring the black box*. New York: Cambridge University Press.
Sako, M. (1999). From individual skills to organizational capability in Japan. *Oxford Review of Economic Policy*, *15*(1), 114–126.
Schroeder, R. G., Bates, K. A., & Junttila, M. (2002). A resource-based view of manufacturing strategy and the relationship to manufacturing performance. *Strategic Management Journal*, *23*(2), 105–117.
Sobrero, M., & Roberts, L. B. (2002). Strategic management of supplier-manufacturer relations in new product development. *Research Policy*, *31*(1), 159–182.
Stein, J. C. (1995). *Internal capital markets and the competition for corporate resources*. NBER Working Papers 5101. National Bureau of Economic Research, Inc.
Stewart, T. A. (1997). *Intellectual capital: The new wealth of organizations*. New York: Doubleday.
Wagner, S. M., & Bukó, C. (2005). An empirical investigation of knowledge-sharing in networks. *Journal of Supply Chain Management*, *41*(4), 17–31.

Chapter 9

High-Tech Small- and Medium-Sized Enterprises: Methods and Tools for External Knowledge Integration ☆

Jeroen Kraaijenbrink

Introduction

Given the numerous government initiatives in existence improving the transfer of knowledge to high-tech small- and medium-sized enterprises (HTSMEs) appears to be a highly relevant topic (Bougrain & Haudeville, 2002). For example, governments provide subsidies, give training, found knowledge-brokering institutes, websites, and support collaboration between HTSMEs and research institutes (Jetter, Kraaijenbrink, Schröder, & Wijnhoven, 2005). Although government initiatives are undoubtedly helpful in supporting the transfer of knowledge into HTSMEs, they are not the only way to support them. An alternative way to support HTSMEs is by providing them with the Methods and Software Tools (MSTs) they need to identify, acquire, and utilize external knowledge. This process of identifying, acquiring, and utilizing knowledge from their environment is termed external knowledge integration (EKI) in this chapter.

There exists a vast array of MSTs that are potentially useful to support EKI in HTSMEs. In this chapter, methods are considered to be ways of thinking and acting when approaching a problem. Examples of such methods are benchmarking, gap

☆ A condensed selection of this paper has been published as part of Chapter 3 of the book that resulted from this project (Jetter et al., 2005).

New Technology Based Firms in the New Millennium, Volume VII
Edited by R. Oakey, A. Groen, G. Cook and P. van der Sijde

analysis, and internal documentation procedures. Tools are considered the embodiment as of such methods in pieces of software. Examples of tools are data mining software, content management systems, and groupware. As a number of studies show, the usage of these MSTs, and the associated tools, is low among HTSMEs and SMEs in general (Bessant, 1999; Corso, Martini, Pellegrini, & Paolucci, 2003). This is not surprising, since most MSTs have been developed by or for large companies, and most studies of MSTs are in large companies (Binney, 2001; Nissen, Kamel, & Sengupta, 2000; Paton, Goble, & Bechhofer, 2000; Ruggles, 1997). Although perhaps not surprising, it is striking to note that an economically crucial group of companies such as HTSMEs is not using potentially useful MSTs to support their EKI processes. This seems to imply that EKI, which is a crucial process for all HTSMEs, is not used to the extent it could be deployed. It is the purpose of this study to find out to what extent this is indeed the case and if so, why. Furthermore, we will generate ideas on what could be done about it.

Although some previous studies have been conducted on MSTs' usage in small firms, they have focused on a rather narrow set of MSTs, or even on the evaluation of a single MST (e.g., Bessant, 1999; Scherf & Böhm, 2005). No study systematically analyzes the usage of a broader range of MSTs for EKI among HTSMEs. To address such a lacuna in the current literature, this chapter presents the results of such a study. This chapter will answer the following three questions:

1. To what extent are various existing MSTs used by HTSMEs?
2. What is the level of satisfaction with these MSTs among HTSMEs?
3. If applicable, what are the causes for a low usage of MSTs among HTSMEs?

The answers to these three questions will provide a better insight into which MSTs HTSMEs use, which MSTs they do not use, and why this is the case. On the basis of these insights, the chapter provides suggestions for how to increase the use of MSTs among HTSMEs.

The chapter is structured as follows. The next section presents the research methods of a survey that was conducted to answer the above three questions. The Results section provides the results of the survey, and the chapter ends with a conclusion, followed by a discussion.

Research Method

The answers to the three research questions were sought by means of a large-scale online and paper-and-pencil-based questionnaire sent to a stratified randomized sample of 1306 HTSMEs in Germany, Israel, the Netherlands, and Spain. The complete survey concerned a wide range of topics related to EKI and was conducted as part of the European project "Knowledge Integration and Network eXpertise" (KINX). Approximately one-fifth of this survey was reserved for the answering of the earlier three questions of this study.

The Questionnaire

On the basis of existing inventories of MSTs (Bullinger, Wörner, & Prieto, 1997; KLUG, 2002; Sebastiano et al. 2002), MSTs were selected to be covered by the questionnaire. The final questionnaire was based in a list of 15 types of methods and 17 types of software tool that could be used to support the identification, acquisition, and utilization of knowledge (Figures 1 and 2). For each MST, respondents were asked to indicate whether they had used it and, if so, whether they were satisfied with it or not. These questions were posed in a simple yes/no format.

To obtain a better insight in the type of methods and tools that were used, a second question was asked on the degree to which the methods and tools that were used were standard or customized. Respondents were asked to indicate on a 5-point scale ranging from strongly disagree (i.e., 1) to strongly agree (i.e., 5) to what extent they agreed with the following two statements: 1) Most of our methods and software that deal with knowledge are especially developed for our company, or 2) Most of our methods and software that deal with knowledge are specific to our field.

Finally, the questionnaire asked for reasons why respondents did not use MSTs more often. On the basis of a meeting with an expert team of HTSME managers, consultants, and academics (i.e., the KINX consortium) and 33 exploratory interviews with HTSME managers, the following options were chosen: "Not thought about it," "There was no need for it," "I am not aware of any," "There are too many to choose from," "They are too expensive," "There are none suitable," and "It is too complicated to learn to use them." Also, respondents had the opportunity to choose the option "Other reason, namely…." This question was put to each of the

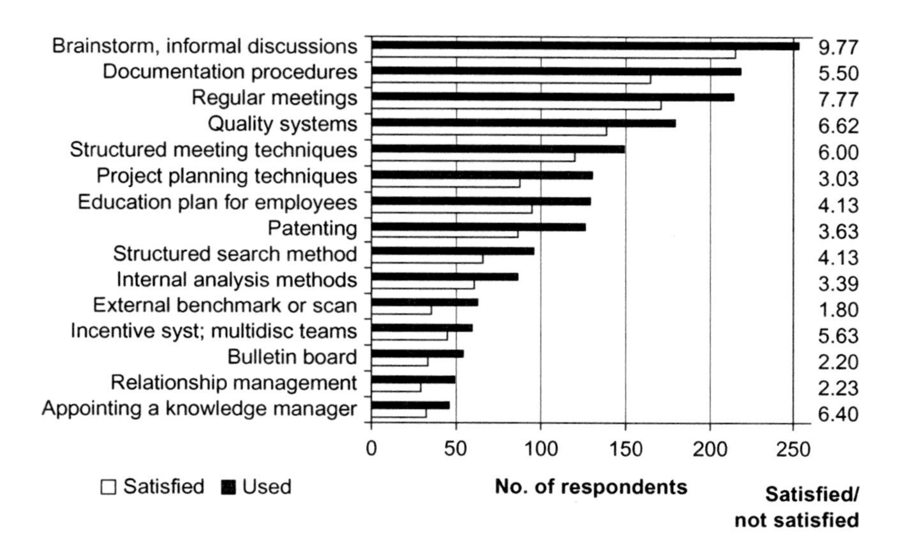

Figure 1: Number of respondents that use certain methods and that are satisfied with them (adopted from Kraaijenbrink, Groen, & Wijnhoven, 2005).

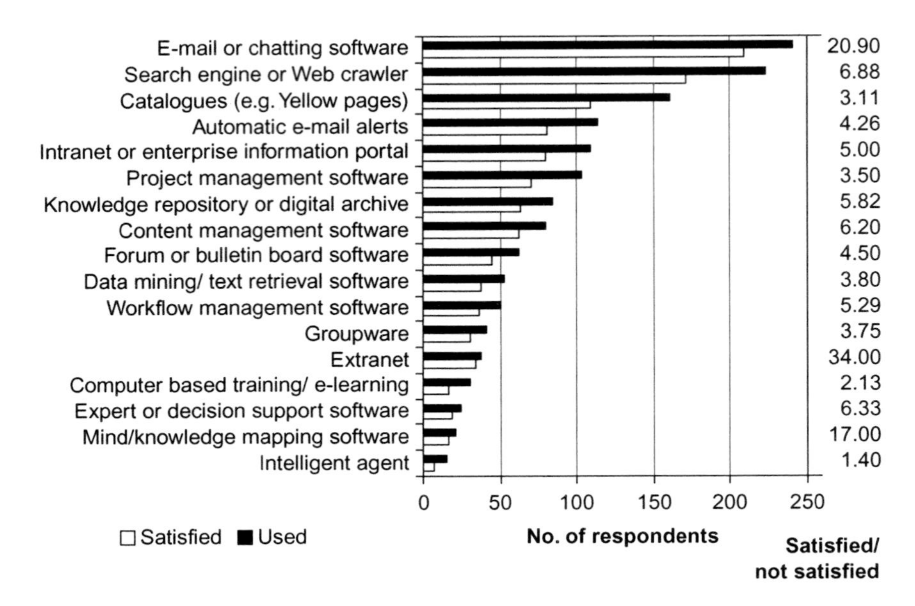

Figure 2: Number of respondents that use a certain software and that are satisfied with it. (adopted from Kraaijenbrink et al., 2005).

three stages of EKI (identification, acquisition, and utilization of knowledge) and for two types of knowledge (customer/market knowledge and technological knowledge).

Sample

To create a sample of HTSMEs, it was necessary to specify what was meant by "SME" and by "high-tech." The formal European definition of SMEs includes companies that have fewer than 250 employees (European Commission, 1996). However, although the average company in Europe has six employees, for those branches that are defined as high-tech manufacturing by the OECD (2001), the average number of employees is 20. This implies that the average size of the companies this study focuses on is more than three times as high as the overall European average. Therefore, companies were included with up to 500 employees in our study. For the definition of "high-tech" companies, we adopted the official International Standard Industrial Classification (ISIC) of high-tech and low-tech industries. This study includes firms of both high-technology and medium-high-technology industries (Table 1).

A major challenge was the selection of high-quality address databases for the questionnaire. Since no database that covered the four countries were not available, four different databases were selected that allowed selection on similar criteria. Because of their high-quality reputations and similarities, the following databases were selected: Hoppenstedt (Germany), D&A HiTech Information Ltd. (Israel),

Table 1: Profile of respondents and their companies.

Industry			**%**	**Year of Foundation**	**%**
24 chemicals and chemical products			10.7	Before 1965	13.1
29 machinery and equipment			28.4	1966–1980	13.1
30 office machinery and computers			11.7	1981–1990	18.0
31 electrical machinery and apparatus			4.1	1991–1995	14.6
32 radio, TV, and communication equipment			19.9	1996–1998	15.5
33 medical, precision, and optical instruments			12.6	1999–2001	16.2
34 motor vehicles, trailers, and semi-trailers			5.0	Missing	9.5
35 other transport equipment			3.2	After 2001 excluded	
Missing			4.4		
Number of employees	**Total**	**On R&D**		**Position of Respondent**	**%**
2–9	14.3%	58.5%		Director/general manager	29.9
10–49	28.7%	23.2%		Manager/head R&D	37.8
50–99	16.5%	5.2%		Manager/head marketing	14.3
⩾100	35.1%	3.4%		Other	12.8
Missing	5.5%	9.8%		Missing	5.2
Mean	89.5	14.8			

National Chamber of Commerce (Netherlands), and AXESOR (Spain). From these databases, a stratified random sample was selected of 1722 HTSMEs. The sample was stratified in terms of country (Germany, Israel, Netherlands, and Spain), size (2–9, 10–49, 50–99, and 100–499 employees), and industry (24 and 29–35 from the ISIC). These companies were contacted by phone and were asked to identify a key informant, who received a personal (web-based or paper-and-pencil-based) questionnaire with an accompanying letter. Although the validity of single-informants research has been debated, Campbell (1955) concludes that this type of sampling can produce results that are valid and generalizable. Also, this chapter concerns with Kumar, Stern, and Anderson (1993) who state that informants were not selected to be representative of the members of a studied organization, but because they were supposedly knowledgeable and willing to communicate about the issue being researched. Since smaller companies are less likely to have such informants than large companies (Mitchell, 1994), companies were allowed to decide themselves who was the most appropriate person to respond. During telephone calls, respondents were asked whether they were indeed the right person in the company to answer the questionnaire. It was expected that this self-selection mechanism would lead to a strong overrepresentation of technology-oriented respondents compared to market-oriented respondents. This expectation was based on an assumption that respondents from HTSMEs would be associated with research and development rather than with marketing. Although our expectation was partly right, a substantial number of market-oriented persons also responded. When the selected respondents did not respond within the indicated period (i.e., two weeks), they were reminded up to two times, which is argued to be the optimal number of reminders (Babby, 1995).

Response

From the 1722 SMEs that were initially selected, 416 firms were excluded from the sample for several reasons, including wrong addresses and wrongly classified as HTSME. A final total of 317 HTSMEs responded, leading to an effective response rate of 24.3%, which is high for a randomized sample of SMEs (Huang, Soutar, & Brown, 2002; Raymond, Julien, & Ramangalahy, 2001).

The profile of the responding companies and individuals is given in Table 1. A comparison (*t*-test and Mann–Whitney test) of respondents with non-respondents showed no significant differences regarding industry type (e.g., a two-tailed significance for *t*-test was 0.904 and for Mann–Whitney was 0.516). However, regarding company size, the difference was significant in that both tests showed significance at the 0.000 level. Companies with 10–49 employees were relatively underrepresented in the responses while companies with over 100 employees were relatively overrepresented. Also, concerning company age, differences were significant (i.e., 0.083 and 0.002 for *t*-test and Mann–Whitney test, respectively). Younger companies were relatively underrepresented, whereas older companies were overrepresented. There was no theoretical reason to assume that these over- and under-representations were relevant to the outcomes of the study. Moreover, a comparison (*t*-test) of early and late respondents on all variables in the complete study showed no significant differences at $p<0.05$ level. Thus, a substantial non-response bias seemed unlikely (Armstrong & Overton, 1977).

Results

The results of the survey are presented in Figures 1–6. Figure 1 presents how many of the 317 respondents used, or had used, one or more of 15 types of methods (i.e., black bars). It also presents how many users were satisfied with each of the 15 methods (white bars). The right column in Figure 1 represents the ratio of respondents that were satisfied with a method and respondents that were not satisfied with that method. Figure 2 is identical to Figure 1, but concerns tools instead of methods.

Figure 1 indicates that the methods that were used most are general methods such as brainstorming (i.e., 253 of 317 = 79.8%), documenting (i.e., 218/317 = 68.8%) and regular meetings (i.e., 214/317 = 67.5%). Figure 1 also shows that the least used methods were appointing a knowledge manager (i.e., 46/317 = 14.5%), relationship management (i.e., 49/317 = 15.5%), and bulletin boards (i.e., 54/317 = 17.0%). This implies, for example, that approximately every sixth to seventh responding company had implemented a bulletin board and appointed a knowledge manager. Although these methods score lowest compared to the other methods, it was found that the number of HTSMEs that have appointed a knowledge manager was remarkably high.

It is evident from the right column in Figure 1 that all ratios of satisfied users and non-satisfied users were above 1. This indicates that, for each of the methods, most users were satisfied. This is not very surprising, since people will use a method most when they are, to some extent, satisfied with it. However, it was realized that the figures include some respondents that have previously used a method, meaning that they are currently not using it anymore. Hence, satisfaction is not as obvious as it seems. As the ratios indicate, there does not seem to be a connection between the number of respondents that use a method and the level of satisfaction: in that the ratios are not higher for methods that are used more often.

When tools were examined, the results are to a large extent similar to those of methods (Figure 2). Again it is general tools that are used most widely including e-mail and chatting (i.e., 241/317 = 76.0%), search engines and web crawlers (i.e., 223/317 = 70.3%), and catalogs (i.e., 161/317 = 50.8%). The relatively large 20% difference between search engines and catalogs implies that e-mail and search engines are clearly the most widely used tools. At the bottom of Figure 2, it is evident that the least used tools were intelligent agents (i.e., 15/317 = 4.7%), mind mapping software (i.e., 21/317 = 6.6%), and expert and decision support software (i.e., 25/317 = 7.9%).

As for the methods, the ratios of satisfied and non-satisfied users were all above 1. High satisfaction rates are notable for e-mail, the extranet, and mind/knowledge mapping software. These latter two are particularly interesting since they are not widely used. Rather, it seems that they are used by a selective group of satisfied users.

Additional useful observations can be made when Figures 1 and 2 are compared. The figures illustrate that methods are used more frequently (i.e., 1855 in total) than tools (i.e., 1451). It is likely that, in practice, the difference is probably even higher than these number express since more types of tools (i.e., 17) than types of methods (i.e., 15) were included in the survey. Thus, although there was a bias in the survey likely to produce a higher score on tools, the score on methods was, in fact, highest.

Another difference is the larger spread of methods that were used compared to the frequent use of only a small set of tools. For example, the three best scoring methods add up to 685 of 1855 = 36.9% of total usage, whereas for tools this number is 625 of 1451 = 43.1%. It is also apparent that this curve in Figure 2 is steeper when compared to Figure 1.

When comparing the satisfaction ratios for methods and tools, it appears that the variation in ratios for methods is less than that for tools. For methods the values lie between 1.80 and 9.77, with an average of 4.83, and for tools the values lie between 1.40 and 20.90, with an average of 5.48. This implies that, although there are some tools that score extraordinary highly on satisfaction, this is not the case for methods. It also implies that, on average, one-sixth of the users were not satisfied.

The results for the questions on to what extent most of the MSTs of a company were specifically developed for the company or for the field are presented in Figure 3. This figure clearly indicates that the MSTs that HTSMEs use were predominantly not specifically developed for them or for their types of companies. This observation

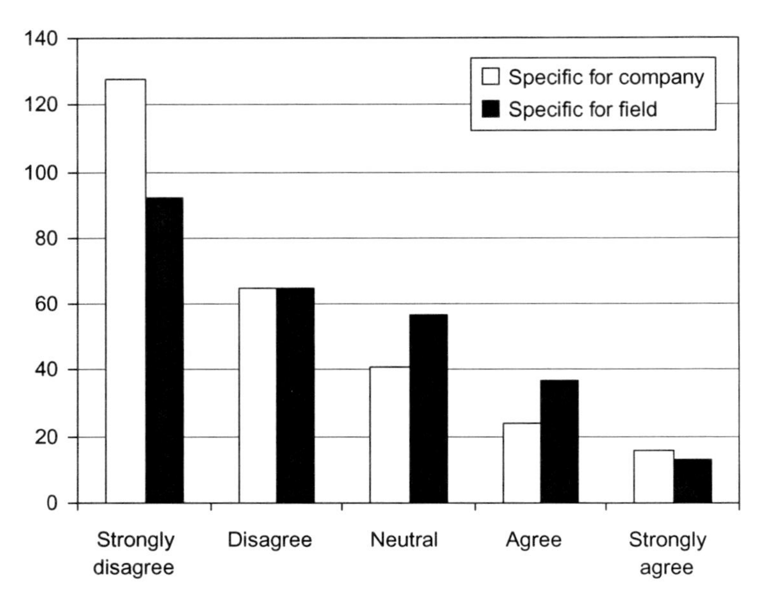

Figure 3: Extent to which methods and tools are specifically developed for company or field.

is consistent with the results of Figures1 and 2 where general types of methods and tools were most used.

As mentioned in Research Method section, respondents were also asked why they did not use more specific methods and software for the identification, acquisition, and utilization of external customer/market knowledge and technological knowledge. The results for these questions are presented in Figures 4–6.

From Figures 4–6 the following results can be derived. First, it is apparent that "Not being aware" is the most important reason for not using specific MSTs, regardless of EKI stage and type of knowledge. The figures also show that "There are too many" and "They are too complicated" are the least important reasons, regardless of EKI stage and type of knowledge. Another observation is that "Too expensive" scores higher than "Not suitable," regardless of stage and type of knowledge. This indicates that, for HTSMEs, the price of MSTs is a larger barrier to MST usage than the suitability of these MSTs.

Moving from identification, through acquisition, to utilization, it can be seen that the reason "No need for it" increases in importance, compared to the other reasons. This indicates that HTSMEs perceive a higher need for MSTs for identification than for acquisition and utilization.

Concerning the category "Other reasons," the reasons that were given most frequently were "No time" and "Company too small and/or specific," which for both could mean as "No need for it" or "There are none suitable."

In general, it can be seen that, when moving from identification, through acquisition, to utilization, the numbers decrease. However, it is probable that this has

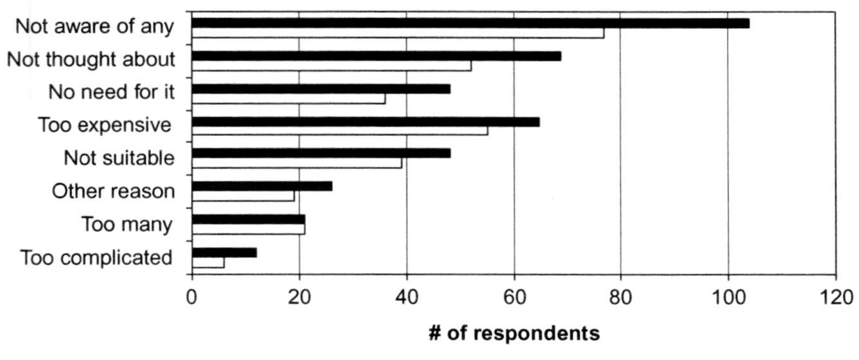

Figure 4: Reasons for not using methods and tools more often for knowledge identification.

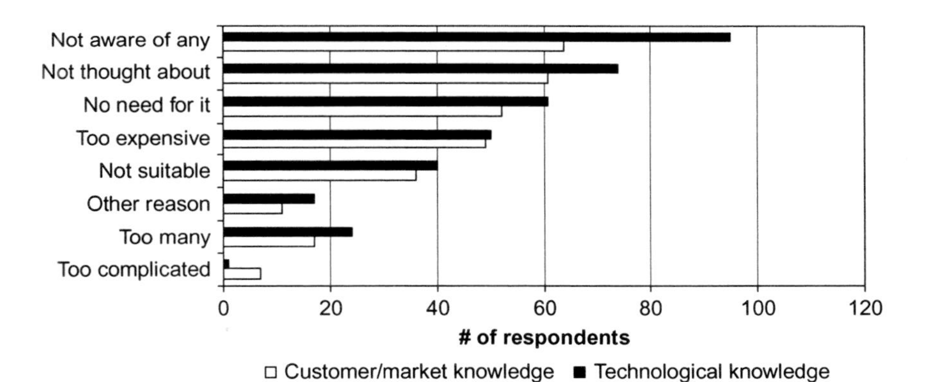

Figure 5: Reasons for not using methods and tools more often for knowledge acquisition.

more to do with a decrease in responses because of repetitive questions than with any other reason. Also, the fact that the numbers for customer/market knowledge are lower than for technological knowledge have to do with the fact that more respondents filled in the questionnaire for technological knowledge than for customer/market knowledge (i.e., respondents could choose, based on their expertise).

Conclusions

This chapter began with three related research questions: 1) To what extent are various existing MSTs used by HTSMEs? 2) What is the level of satisfaction with these MSTs among HTSMEs? 3) What are the causes for a low usage of MSTs among HTSMEs?

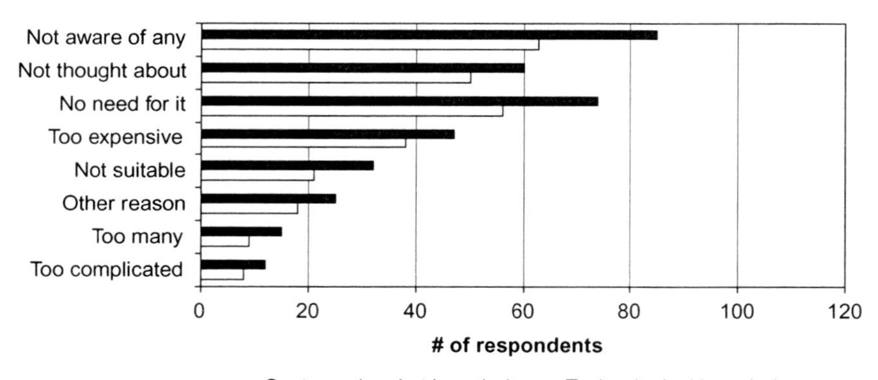

Figure 6: Reasons for not using methods and tools more often for knowledge utilization.

Concerning the first question, the results have shown that the extent to which MSTs for EKI are used by HTSMEs varies substantially, dependent on the particular MST. Figures 1 and 2 indicate that the usage of general, relatively low-profile methods and tools is high, whereas the use of more specific and complicated MSTs is low. Also, the figures show that methods are used more than tools and that the MSTs they use are usually not developed especially for their firm or field. This is also what might be expected from HTSMEs, as they usually do not have the money or expertise available for investing in highly specialized tools or methods. Compared to existing research on HTSMEs use of MSTs, this answer to the first research question refines the general observation that HTSMEs hardly use MSTs for EKI. Also, it is likely that the usage of methods is in general not as low as might have been expected. For example, Figure 1 shows that approximately every seventh company has appointed a knowledge manager. This is a very high number for such an expensive measure. Nevertheless, the results do confirm the observation that specific tools and methods are hardly used by HTSMEs.

The second question concerned the satisfaction level of MSTs. From Figures 1 and 2 it can be concluded that, in general, most users are satisfied with the MSTs they use, or have used. However, on average, approximately one-sixth of the users of an MST are not satisfied. From these figures it can also be concluded that higher usage is not associated with higher satisfaction. Rather, there seems to be no connection between the usage and satisfaction rates. Finally, there are some tools where an extraordinary large share of users are satisfied (i.e., e-mail and chatting, mind mapping, and extranets).

With respect to the final research question, the results in Figures 3–6 show that "Not being aware of MSTs" is the most important reason for not using them, followed by "Not thought about" and "No need for it." Reasons for not using MSTs that were hardly mentioned are that the MSTs were too complicated, or not suitable for HTSMEs, or that there were too many MSTs to choose from.

Discussion

The final section of this chapter discusses the implications of the answers given to the three research questions for research and practice.

For practice, the main implication of these results concerns the question how the usage of MSTs for EKI can be increased among HTSMEs. A question that should precede this question is whether increased usage is desirable. Considering the answers given to the three research questions, this second question should be answered affirmatively. The results show that most users are satisfied and that the most important reasons for not using MSTs is that companies are not aware of any, or have not thought about them. Hence, it appears that the low usage of MSTs amongst HTSMEs is not caused by some general defect in MSTs, but more by a lack of awareness and knowledge among HTSMEs employees.

To increase the usage of MSTs by HTSMEs, it is thus important to increase HTSME employees' awareness of MSTs, and to improve the accessibility and publicity of MSTs. One potentially fruitful way to do this is by means of an Internet portal. Such portal could aggregate and give access to a large number of MSTs and their suppliers. It can even provide a diagnosis of problems and a matching of MSTs with these problems.

However, since the Internet is a passive medium, developing an Internet portal alone is not sufficient. Additionally, it is necessary to create active and targeted communications with HTSMEs to make them aware of the existence of such a portal, or more generally, to make them more generally aware of the existence of potentially useful MSTs. As they are generally well-known and well-connected to HTSMEs, SME agencies, innovation centers, and Industry Associations seem to be among the most appropriate institutions to deliver such information. Should communication not be sufficient, such institutions could also provide training that render managers of HTSMEs better aware of EKI as an important process and of the opportunities that MSTs provide to support this process.

Given the relatively high satisfaction rates of those HTSMEs that use particular MSTs, improving the quality of the MSTs themselves seems of less importance. Since price was mentioned to be a more significant barrier for the use of MSTs than suitability, it seems more important to lower the prices of existing MSTs. This does not only include the price of purchasing the right to use a particular MST but also the price of implementing, maintaining, and using it. Although suppliers of MSTs have of course a major role to play in this, it is expected that the institutions mentioned above can also play an important role here. For example, as representatives of HTSMEs, such institutions should be able to negotiate with suppliers of MSTs, possibly arranging collective licences for a number of HTSMEs together.

In addition to these implications for practice, this chapter has also implications for further research. The chapter has not only answered some questions, but it has also generated other new questions. As the current study has made a broad inventory of the usage of MSTs by HTSMEs, it seems useful that further research might consist of in-depth analyses of MSTs usage and non-usage. Thus, future research should have a closer look at the MSTs that have a high usage rate and MSTs that have a high satisfaction rate. Future questions that need answers are: what makes these MSTs so good or suitable, why are they used, and how are they different from MSTs with low usage and satisfaction rates?

Also, future research should further investigate the type of companies that use and are satisfied with particular MSTs. It might not be so much the type of MST that explains usage, but more the type of company that uses the MST. To investigate this, a comparison should be made between, for example, companies of different sizes, ages, and high-technology industries.

Finally, since it appears that satisfaction and usage do not seem to be closely related, the question arises as to what does explain usage? Some hints have been obtained from the reasons for not using MSTs, but further research is needed here.

Acknowledgement

The research done in this study was financially supported by the European project "Knowledge Integration and Network eXpertise" (KINX), No. G1RD-CT-2002-00700.

References

Armstrong, J. S., & Overton, T. S. (1977). Estimating nonresponse bias in mail surveys. *Journal of Marketing Research*, *14*(August), 396–402.

Babby, E. (1995). *The practice of social research* (7th ed.). Belmont, USA: Wadsworth Publishing Company etc.

Bessant, J. (1999). The rise and fall of "supernet": A case study of technology transfer policy for smaller firms. *Research Policy*, *28*, 601–614.

Binney, D. (2001). The knowledge management spectrum-understanding the KM landscape. *Journal of Knowledge Management*, *5*(1), 33–42.

Bougrain, F., & Haudeville, B. (2002). Innovation, collaboration and SMEs internal research capacities. *Research Policy*, *31*, 735–747.

Bullinger, H. J., Wörner, K., & Prieto, J. (1997). *Wissensmanagement Heute: Daten, Fakten, Trends*. Stuttgart: Fraunhofer Institut für Arbeitswirtschaft und Organisation (IAO).

Campbell, D. T. (1955). The informant in quantitative research. *American Journal of Sociology*, *60*(4), 339–342.

Corso, M., Martini, A., Pellegrini, L., & Paolucci, E. (2003). Technological and organizational tools for knowledge management: In search of configurations. *Small Business Economics*, *21*(4), 397–408.

European Commission. (1996). Commission recommendation of 3 April 1996 concerning the definition of small and medium-sized enterprises. *Official Journal of the European Communities*, *L107*, 4–9.

Huang, X., Soutar, G. N., & Brown, A. (2002). New product development processes in small and medium-sized enterprises: Some Australian evidence. *Journal of Small Business Management*, *40*(1), 27–42.

Jetter, A., Kraaijenbrink, J., Schröder, H.-H., & Wijnhoven, F. (2005). *Knowledge integration: The practice of knowledge management in small and medium enterprises*. Heidelberg: Physica-Verlag.

KLUG. (2002). *Klug: "Kenntnisse leiten zu unternehmerischem Gewinn"*. Avialable at http://www.iw-klug.de

Kraaijenbrink, J., Groen, A., & Wijnhoven, F. (2005). Knowledge integration by SMEs-practice. In: A. Jetter, J. Kraaijenbrink, H.-H. Schröder & F. Wijnhoven (Eds), *Knowledge integration: The practice of knowledge management in small and medium enterprises* (pp. 29–46). Heidelberg: Physica-Verlag.

Kumar, N., Stern, L. W., & Anderson, J. C. (1993). Conducting interorganizational research using key informants. *Academy of Management Journal*, *36*(6), 1633–1651.

Mitchell, V.-W. (1994). Using industrial key informants: Some guidelines. *Journal of the Market Research Society*, *36*(2), 139–144.

Nissen, M., Kamel, M., & Sengupta, K. (2000). Integrated analysis of design of knowledge systems and processes. *Information Resources Management Journal*, *13*(1), 24–43.

OECD. (2001). *Science, technology and industry scoreboard 2001: Towards a knowledge-based economy* (e-Book). Paris: OECD Publishing.

Paton, N. W., Goble, C. A., & Bechhofer, S. (2000). Knowledge based information integration systems. *Information and Software Technology*, *42*, 299–312.

Raymond, L., Julien, P.-A., & Ramangalahy, C. (2001). Technological scanning by small Canadian manufacturers. *Journal of Small Business Management*, *39*(2), 123–138.

Ruggles, R. (1997). *Knowledge tools: Using technology to manage knowledge better*. Cambridge, MA: Ernst & Young LLP.

Scherf, A., & Böhm, K. (2005). KnowBiT-knowledge management in the biotechnology industry. *Lecture Notes on Artificial Intelligence*, *3782*, 721–728.

Sebastiano, G., Coviello, A., Garavelli, C., Gorgoglione, M., Kemp, J., Moult, D. L., & Bredehorst, B. (2002). *Framework and guidelines for human factors change management in KM*. IST Project No 2000-26393.

Chapter 10

The Relevance of Disruptive Technologies to Industrial Growth: A Conceptual Critique

Raymond P. Oakey

The Key Role of Technological Change in Industrial Growth

Since the onset of the industrial revolution in England during the late 18th century, it has become increasingly clear how advances in technology have played a pivotal role in delivering wealth-creating economic growth, ranging from major advances in the generation of industrial power, initially through steam engines (e.g. successively by Nucomen, Watt and Trevithick), to the design of labour saving industrial machinery and working practices (Smith, 1776; Marx, 1867; Solow, 1957; Denison, 1967; Mansfield, 1968; Freeman, 1982). These advances have not merely resulted in industrial progress but have triggered changes in industrial location (e.g. water powered to coalfield sites in the cotton textile industry) (Riley, 1973), dictated population distributions and fixed the positions of major industrial cities within national and world regions. Indeed, perhaps, the most ambitious attempt to establish the major impact of revolutionary technological change on macro-level industrial performance was the explanation by Schumpeter of Kondratiev's 'long wave' industrial cycles (Kondratiev, 1925) in which upswings in world economic activity were linked to the introduction of pervasive new technologies caused by their ability to reduce unit prices, increase efficiency and be broadly applicable across large sectors of industry (e.g. stream and electric power) (Schumpeter, 1939; Freeman, 1986).

However, the relatively recent impacts of technological progress can be set within a much longer time frame in which progress, although intermittent and erratic in the short term (and often punctuated by periods of stasis or even regression, e.g., the fall of the technically advanced Greek and Roman Empires and the economic activity sapping spread of bubonic plague in 13th-century Europe), has been generally persistent and beneficial over the longer term. Moreover, if this very broad historical

New Technology Based Firms in the New Millennium, Volume VII
Edited by R. Oakey, A. Groen, G. Cook and P. van der Sijde

perspective is taken, it is relevant to the following main argument of this chapter to observe that progress has often been seen as both 'disruptive' and unwelcome to those whom, in the short term, were disadvantaged by technological change. Yet, in the medium to long term, uncomfortable short-term radical changes, which have often triggered social unrest (e.g. the Luddite riots), have proven to be both welcome and inevitable. Thus, as many experts on technological change (cited above and below) have concluded, advances in technology are inevitable and ultimately welcome ramifications of economic competition.

Nonetheless, the acknowledgement by mainstream economics of the importance of technological change to economic growth has never, in the past, been a wholehearted process, in conditions where the critical but *complex* impacts of technological progress on industrial growth do not fit easily into relatively simple economic equations designed to model the performance of economies, usually by means of government statistics collected for a different purpose (e.g. R&D expenditure; R&D employment and patent registration) (Oakey, Rothwell, & Cooper, 1988), deployed at various spatial scales (Grilliches, 1957; Thwaites, 1978; Porter, 1998). Moreover, although many undergraduate-level economics textbooks have traditionally tended to give technological change insufficient weight (e.g. Marris, 1964; Dorfman, 1972), a growing number of western economists have struggled to measure the clearly critical (but difficult to isolate) impacts of technological change on industrial performance at national levels in the latter part of the 20th century (Solow, 1957; Schmookler, 1966; Dension, 1967; Mansfield, 1968; Freeman, 1982).

Thus, while some debate remains regarding the accuracy involved in measuring the *total* impact of the introduction of significant new technologies on industrial growth, particularly regarding its extent (e.g. the financial benefit of microprocessors as both products and processes since 1970), a central argument of this chapter is that very few experts on economic performance, at the regional or national levels (Thwaites, 1978; Oakey, Thwaites, & Nash, 1979), now question the assertion that such impacts are *substantial* and that radical and recurrent technological changes are inevitable and essential. This imperative is confirmed by the emphasis that governments, universities and industrial firms place on overtly attempting to make progress into new and emerging areas of technology derived from basic academic and industrial scientific research and use these advances to stimulate regional and national economic performance. These much sought after 'blue sky' discoveries have caused, and will increasingly provide, the basis for new industries, often geographically clustered locations (Oakey, 1985; Swann, Prevezer, & Stout, 1998; Oakey, Kipling, & Wildgust, 2001; Porter, 1998). Moreover, given the rapid progress of a number of developing nations over the past 30 years, mainly in the far east, (e.g. India, China, Taiwan and South Korea) to dominate mature 'end of cycle' and even 'mid-cycle' industrial technologies (e.g. textiles, motor vehicles, consumer electronics and computers), the governments of most developed economies view the pursuit of radical new technologies as a means of preserving their high wage economies by concentrating on 'start of cycle' technologies (e.g. biotechnology, artificial intelligence, nanotechnology, space science and multimedia).

Moreover, risky new technological inventions have been seen as a way for developed economies to creating technological solutions that render developed economies yet more rewarding for the sophisticated consumer (e.g. new medical cures, new ways of communicating and new methods of travel). Expressed in terms of a 'SWOT Analysis', technological progress simultaneously might be seen as *both* an 'opportunity' and a 'threat'. This chapter will fundamentally argue that, since technological change is generally relentless and inevitable under conditions of free market competition, from the viewpoint of technological disruption within economies, it is stasis that is disruptive, since it arrests an otherwise virtuous process. Change only becomes a threat if opportunities for progress are not taken which, in a competitive economic environment, would be a very dangerous strategic approach for an individual firm, a regional economy or a nation, and is an approach that would be bound to fail in the medium to long term (Rothwell & Zegveld, 1981). Although the pursuit of technological progress is both an essential and natural objective for mankind, it is only *necessary* but not sufficient for improvement to occur. Sufficiency is gained by attempting, *and successfully achieving*, technological progress, although this is a risky strategic option that can turn opportunity into a threat, through fruitless high spending on R&D and marketing, when unsuccessful.

The Conflict between Step Change and Smooth Curve Models of Industrial Growth

Recently, there has emerged a tendency for researchers engaged in work on technological change to use the term 'disruptive technologies' when discussing the manner in which radical new technologies upset any given technological *status quo*, a term generally attributed to Christensen (1997). The use of this rather negative term is intriguing since, in most contexts, it would imply that a previously virtuous regime has been interrupted by an unwarranted aberration, similar to the manner in which a flight programme of an airport might be 'disrupted' by bad weather.

This chapter will question the tenor of such a term by suggesting that, in most cases involving technological progress, such 'disruptive' acts, at worst, revitalise the technological *status quo* with a better rate and trajectory of advancement for existing core technologies and, at best, *replace* the *status quo* with a new technological solution, often derived from a completely different (possibly new area) of science (e.g. the replacement of conventional mail with Internet e-mail communication). However, technical advances, although desirable in theory, are often blocked by various forms of inertia most importantly including the monopolistic (or oligopolistic) behaviour of large firms in circumstances where one (or a small number of) producers deliberately *arrest* technological progress due to their ability to enjoy 'super profits' from existing sub-optimal technologies steeped in extensive previous (inertia causing) investment (Geer, 2003).

The impetus to describe technological progress as 'disruptive', to some extent, may stem from a natural individual (and group) human dislike of the uncertainty engendered by surprise events (in this case technological revolutions). However, in most cases where technology is concerned, we have an illogical desire *both* for stability and for the fruits of rapid technological progress in circumstances where we, for example, value the benefits that mobile telephones bring but complain when others use them on public transport. From a psychological viewpoint, there appears to be a 'glass half full or half empty' problem in that we often view technological change rather schizophrenically as *both* desirable and damaging, depending on whether we are using the phone or being irritated by its use. However, while such ambivalence might be understandable under conditions of short-term uncertainty, and easily explained by the work of behavioural theorists as sub-optimality, in circumstances where the crude maximisation of monetary rewards is traded for various forms of 'psychic income' that might illogically conflict with wage maximisation (e.g. easy communication, privacy, job satisfaction and free-time) (Simon, 1955; Cyert & March, 1963; Pred, 1965), the central argument of this chapter will be that change is unavoidable and *normal* and is increasing in pace. Indeed, it is stasis that is a more often a disruptive force that inhibits economic, and by implication, human progress.

Reasons for 'Disruption'

Most of the evidence that exists on technological change suggests that, even in a perfectly responsive market where no monopolistic distortions exist, it is not easy for new technologies to replace (or disrupt) the *status quo* (Schumpeter, 1943; Kamien & Swartz, 1983; Oakey, 1993). The public are discerning and are not easily 'duped' by the glamorous advertising of products, as witnessed by the original massive success of mobile phone technology, and the subsequent relative slow uptake of its 3G 'next generation' replacement. The consumer has a very well-developed sense of what has high and widespread personal utility (e.g. the basic mobile phone) and technological 'improvements' that are, by comparison, 'cosmetic' (e.g. 3G phones incorporating cameras). Since 'disruptive' technologies will only be strongly disruptive if what they offer is *substantially* better than what already exists, we should not worry that substantive disruption resulting from spurious technological 'advances' will occur by causing the inconvenience of change without benefit.

Freeman (1982) has illustrated this problem succinctly when examining the original work of Kondratiev (1925) on 'Long Wave' theory and Schumpeter's later work that attributed 'upswings' in the world's economic cycle to crucial technological inventions (Schumpeter, 1939). Looking to the future, Freeman has argued that, for any new technology to have broad impact (and be commercially 'disruptive'), it must possess three major attributes; first, an ability to achieve rapid reductions in the cost of its production (say in less than 10 years); second, a *concomitant* achievement of a rapid improvement in performance of the new technology (or specification) (Figure 1); and third, partly as a consequence of points one and two above, the new technology should have a wide impact, not only by creating a new type of industrial activity but by reducing operation costs across a wide range of other existing

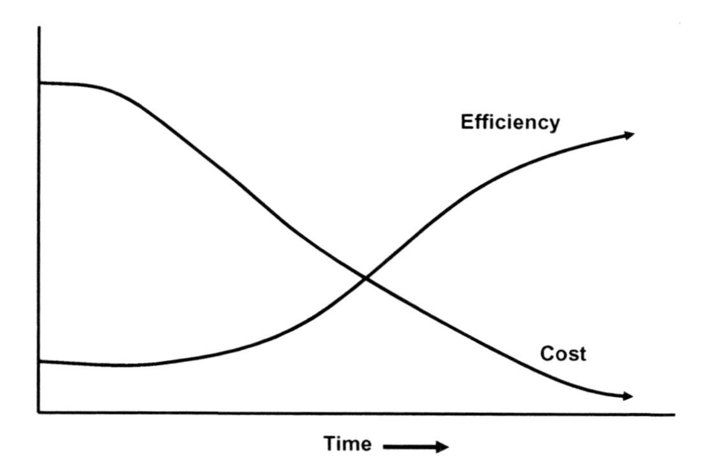

Figure 1: The Interaction between Cost and Efficiency for New Technologies with High Market Penetration Potential.

industrial sectors. Freeman argues, with both strong theoretical logic and widespread evidence from the industry, that evolving semiconductor technology has adequately fulfilled the three above requirements and has produced massive impacts in the factory (e.g. robots) and the office (e.g. word processing). While these achievements have 'disrupted' (and virtually destroyed) some industrial activities (e.g. thermionic valve and mainframe computer production), the overall net gain to the world economy has been undeniable (Freeman, 1986).

Philosophical Perspectives on How New Knowledge is Created and Absorbed

There are many parallels between the manner in which scientific knowledge is generated and absorbed in academia and the manner in which industry utilises new technology. In this context, the use of the term 'paradigm', defined as a regime of knowledge, has often been used in both academic and industrial contexts (Dosi, 1993). Arguments in philosophy in general, and the philosophy of science literature in particular, often hinge on semantic interpretations. Perhaps, one of the great arguments in the philosophy of science occurred between Karl Popper (1965) and Thomas Kuhn (1962), in which both authors made some telling contributions regarding the manner in which scientific knowledge evolves over time. These views are of concern to this chapter, both because, as noted above, the principle of 'dominant paradigm' is useful in describing knowledge generation in both academia and industry, and also simply because there is a practical link in that the results of scientific discoveries often form the basis for new industrial products (or whole industries). Thus, the contrasting arguments of Popper and Kuhn, on how science progresses, can be directly used to discuss how industrial technological progress occurs.

In general, Popper's argument that 'progress' (implying improvement) is a more useful and accurate description of what science and industry achieve than Kuhn's more limited proposition on 'change', since in areas such as medicine and manufacturing, we can point to examples of real progress (e.g. the invention of anaesthetics and the semiconductor) (Magee, 1973; Harvey, 1973). Nonetheless, intriguingly, Kuhn's conceptions on 'normal science' and 'paradigm shift' are equally useful and are often a more realistic explanation of the rate and the manner in which knowledge is assimilated.

Figure 2 compares Popper's view of progress with those of Kuhn to show how 'progress' might be achieved, either steadily over time (dotted straight line) according to Popper, or changes by means of erratic 'steps' (solid lines), according to Kuhn, in which periods of 'normal science' are punctuated by sudden 'paradigm shifts', when revolutionary progress is achieved almost 'overnight', and old regimes are replaced by 'young Turks'. Kuhn adds that such change in science within academic institutions is often of a generational nature (Kuhn, 1962).

However, as noted above, although these concepts were developed to explain changes in pure science, they have equal utility in explaining how change can occur in industrial contexts in circumstances where institutional cliques, represented by universities, can be replaced by the monopolistic tendencies of large well-established industrial firms. In both the academic and the industrial instances, the *status quo* is often maintained to serve the narrow imperatives of these power elites, against the general interests of the public. However, it is significant to the main arguments of this chapter that, in an industrial context, it is generally the brief periods of paradigm shift that produce progress, while 'normal' periods, although comfortable, do not deliver benefits. Thus, it is strongly argued here that it is through periods of 'disruption' that progress is achieved and that 'disruption' is the healthy *norm* for industrial behaviour and by implication that stasis is an aberration. This observation fits well with the view of Schumpeter, discussed below, in which 'waves of creative destruction' were seen as key deliverers of radical technological progress (Schumpeter, 1939).

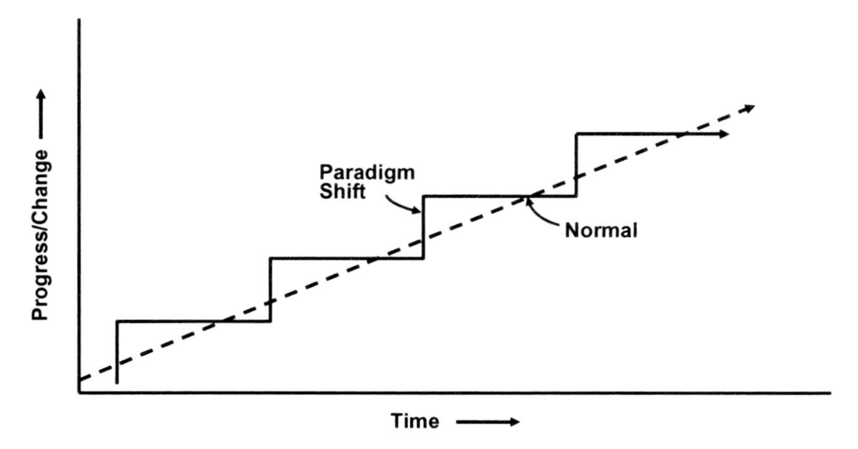

Figure 2: Gradual versus revolutionary conceptualisations of scientific progress.

Another means of depicting progress in an industrial context is to view the principles expressed in Figure 2 in terms of sequential product life cycles (or sigmoidal growth curves) in which a maturing 'existing technology' is supplanted (or disrupted) by a radical new and more efficient replacement technology (Figure 3). Significantly, although the 'destruction or disruption' of one product by another could be seen as a traumatic event (often causing a collapse in the original product's sales and a subsequent loss of jobs), Figure 3 implies that, in many cases, there is a strong evolutionary dimension to the process of technological progress, that is likely to bring major medium- to long-term benefits to the consumer in circumstances where an individual might be damaged by the technological progress (i.e. made redundant), while benefiting from such an advance as a customer.

The diagram depicts the relationship between two product life cycles in terms of efficiency/sales over time (as only part of a longer evolutionary trend). For product 1, basic research creates the new technology on which the product is based, while applied research delivers the most rapid period of efficiency gains as the new technology involved is honed, and performance and production problems are eliminated. Subsequently, development R&D is mainly concerned with 'fine tuning' of the product design and performance and is largely 'cosmetic', since little efficiency gains are achievable at this later stage due to the 'law of diminishing returns' setting in during the latter stages of any product life cycle.

At the point where P1 has reached the stage of only marginal gains in efficiency at the top of the curve, a number of possibilities exist. First, in some cases, due to arriving at a technological 'brick wall', progress is not possible for technical reasons (at least for some time), and P2 does not come 'on stream' to replace P1 (e.g. the lack

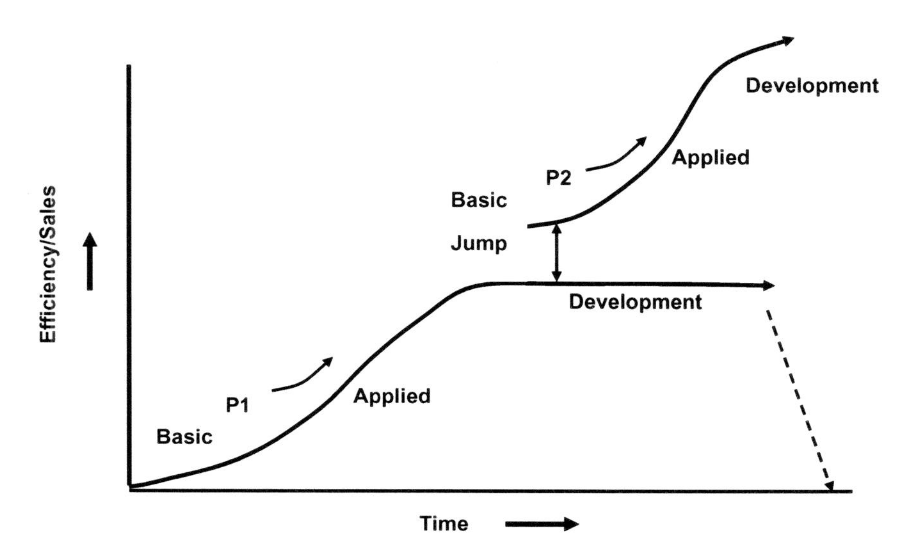

Figure 3: The interruption of P1 by P2 thorough a "jump" in efficiency of solving a common problem.

of an early replacement for Concorde). Second, due to monopoly (or virtual monopoly power when large firms act in consort), in circumstances where a substantial proportion of the industry level output is dominated by one firm (or a small number of firms operating oligopolistically), there is no impetus for manufacturers to produce a replacement to P1, due to the inertia caused by past investment in P1 (i.e. production equipment, sales facilities and servicing — e.g. motor vehicle production based on the internal combustion engine), the high profitability of the *status quo* and lack of competition. Significantly, in such circumstances, the interests of the customer can be neglected. Looked at from the theoretical perspective of the conflict between 'demand pull' and 'innovation push' as a motive for technological change, 'innovation push' from the manufacturer tends to be negated by the highly seductive *status quo* position in which high profits and low uncertainty are enjoyed at minimal R&D cost, while demand pull is also curtailed by the inability of the public to obtain better performing products under condition of monopoly (or oligopolistic) power in which market and technology 'fixing' is achieved by a single or small group of producers, acting in concert (Oakey, 1993; Geer, 2003).

There are two major ways in which this hiatus position of a 'log jam' in technological progress can be broken to restore a more normal and healthy position of general technological competition. First, the market can 'free itself' by actors with no stake in the monopoly *status quo* breaking into the market with an alternative technological solution to restore competition, often by new small firm entrepreneurs (this option will be considered in detail in the section 'The Key Role of the Entrepreneur' below). Second, governments may restore the competitive equilibrium, either through antimonopoly legislation (common in Western developed economies) or through entering into the market as a customer/manufacturer, to promote the development and production of a particular technology determined to be of economic, military or a combined strategic importance to the nation (e.g. Airbus Industries).

If we return to Figure 3, this diagram can be used to illustrate a 'real life' example of how government 'demand pull' spending has provoked a much needed new technical solution in which P1 was superseded by a much better technical solution in the form of P2. A major impediment to the development of early computers was the use of thermionic valves to provide the electronic switches necessary for the construction of computer functions. Because, to improve the memory and operating functions of computers, thousands of switches were required, the early computers were huge, with banks of valves that filled large rooms. There was intense demand from the military establishment, especially in the United States, to find a new type of electronic switch to replace the thermionic valve that was cheaper to produce, more robust and crucially much smaller in order that computer memory power in particular could be rapidly improved (Morse, 1976; Freeman, 1982; Oakey, 1984; Cardullo, 1999).

If this example is imposed on Figure 3, P1 equates to the thermionic valve in circumstances where the 'flat' (small residual progress) development stage of the product development curve was reached by about 1950. An inability to further develop the thermionic valve led to intense *demand pull* from the US military establishment to produce an alternative, a motive that, for 20 years, between 1950 and 1970, was successively driven by the Korean War, the Cold War and The Space Race (Rothwell & Zegveld, 1982). The government funding of research aimed at improving of the thermionic valve throughout this period as a means of increasing the power of computers used to control aircraft, missiles and eventually the space shuttle, initially prompted development of the transistor in the late 1940s (P2 in Figure 3), and through progressive miniaturisation, integrated circuits and the microprocessor family of products in the late 1960s.

The replacement of the old technology by the new, as Figure 3 depicts, afforded an immediate 'jump' in efficiency, which was subsequently augmented by the fact that this new solution was only at the *beginning* of a new product development curve where gains for a given input of R&D investment throughout the 'applied' phase of product development (Figure 3) would be rapid during the steep part of the development curve. Moreover, the miniaturisation of silicon-based electronics, unlike many defence-induced technologies, found widespread use in terms of civil applications and, as noted above, has led to efficiency gains in many existing forms of production (e.g. robots in the motor vehicle industry) and has been the enabling core technology that has allowed new products to emerge (e.g. word processors, mobile phones and the Internet) (Freeman, 1986). This was a case where government defence spending 'demand pull' triggered a radical 'breakthrough' invention that heralded a generally 'smooth' efficiency development of the 'electronic switch' function through semiconductors, integrated circuits and microprocessors between 1945 and 1970.

However, from the broader perspective of this chapter, the above example indicates that, although the arrival of P2 is highly 'disruptive' to sales of P1, the far less desirable option of the non-disruption of P1 (i.e. in this case, a continuance of the inefficient thermionic valve option) would be a much greater *disruption* of technological progress; thus confirming, as argued above, that periods of stasis in terms of the process of technological change are the aberration and not the process of change. Moreover, it is also probable that, without the demand-pull intervention of the US government over 35 years, competition between large existing electronics firms, steeped in thermionic valve technology, would not have produced such substantial progress.

Thus, to draw together the arguments of the previous paragraphs, Figures 2 and 3 may be amalgamated in that the step changes of Figure 2 can be broadly superimposed on the product cycles of Figure 3 (Figure 4). Here, the periods of rapid progress (i.e. the applied stage of the product life cycle in Figure 3) can be broadly equated with the 'paradigm shift' step change of Figure 2. Clearly, the curved line of product efficiency growth in Figure 3 is less abrupt than the vertical line of efficiency growth in Figure 2. However, it should be noted that while the step change concept is a *simplification* of reality to aid explanation, it is also true that, in many fast moving high technology sectors, complete product life cycles last less than 5 years (Oakey, 1984, 1995), implying a 'blip' shape to the product life cycle that both rapid growth *and* decline occurs, with almost no development phase, over a very short life cycle (Oakey, 1984).

Similarly, the 'normal science' flat section of Figure 2 where little or no efficiency gains occur is directly synonymous with the mature 'development stage' of the

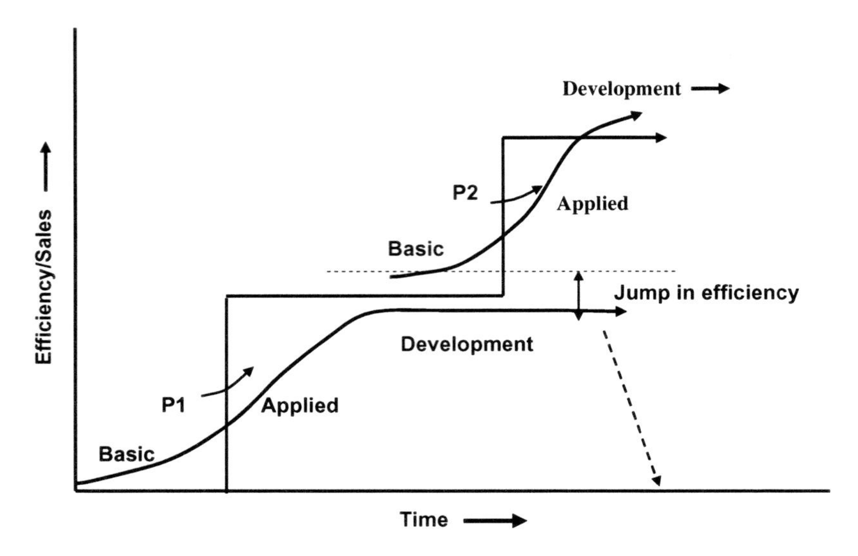

Figure 4: A combination of the concepts displayed in Figures 2 and 3.

product life cycle in Figure 3. As noted above, in both instances, institutional inertia can be proposed as a possible cause of stasis. While Kuhn viewed 'normal' science within academia as a stage in scientific development when institutional inertia was often caused by an 'old guard' of scientists protecting the received wisdom *status quo* (which was often generational in nature), it has been suggested in this chapter that, in an industrial context, monopoly power (or virtual monopoly power involving oligopolistic behaviour) can achieve the same static result. In both instances, the key common point to note is that, although steady progress is clearly generally desirable and less traumatic from *both* producer and consumer viewpoints in most contexts (e.g. in pharmaceuticals), there is often a critical divergence between what is desirable for society (i.e. steady progress) and what might be beneficial for individuals (or groups of individuals [or firms]), which is often preservation of the *status quo*, from which they exclusively benefit.

The Key Role of the Entrepreneur

Although paradoxical, it is often the case that the longer a period of arrested industrial progress continues, the greater the shift that will be caused when a 'revolutionary change' eventually occurs. It is analogous to examples from nature (e.g. earthquakes) that periods of stasis are often not evidence of absolute calm but, more realistically, represent a suppression of 'under the surface' pressure for change, which means that a long period without an earthquake is often an indicator of an eventual disastrous event. It is also perverse in an industrial context that, as pressure for progress builds (e.g. caused by poor product safety or pollution), manufacturers who are 'keeping the lid on', a need for change, often resist new ideas because they are afraid of being swept away by the sudden change, which they know will be both dramatic and inevitable, possibly because they have not performed the necessary R&D that would equip them for a new technological regime, and that they will not survive the revolution. Moreover, from a personal viewpoint, they are often nearing retirement. The much reported difficulties experienced by Frank Whittle in the late 1920s when attempting to sell his new jet engine design to large UK Aircraft manufacturing firms, steeped in piston engine technology, bear witness to the potency of the above-noted phenomenon of resistance, in this case, to an *inevitable* shift away from a piston engine technology paradigm that had, by the mid-1930s, reached its efficiency ceiling. Although Whittle was eventually successful, the delay in developing jet engine technology bears witness to the power of such inertia, sufficient to seriously damage the United Kingdom's Second World War effort.

However, resistance to technological progress by large well-established firms is not a universal phenomenon. For example, in the pharmaceutical industry, large firms commonly invest massive sums of capital into R&D aimed at producing new 'world beating' drugs, in circumstances where their share price can rise or fall rapidly, depending on the optimism of reports from field trials of a new potentially 'world beating' drug under development. Nonetheless, there remain other sectors of industry where large firm dominate in circumstances where necessary technological change (often demanded by social and political pressures) is slow to progress. In sectors as diverse as banking, energy and motor vehicle production, large firms often offer goods and/or services that are worryingly similar in both type and price, under conditions where desirable necessary changes deemed beneficial to the public are slow to appear (e.g. banks [interest rates; loans to small firms]; energy [wind and solar power]; motor vehicles [electric power; fuel cells]). Indeed, desirable changes, which large firms are reluctant to undertake, can often be achieved 'overnight', by governments who unilaterally declare, for heath or other reasons, that a technological change is mandatory (e.g. lead-free petrol engines in the United States during the 1970s) (Rothwell & Zegveld, 1981).

Notwithstanding these variable large firm behaviour patterns, a major mechanism for change provided by the free market system is that of entrepreneurial action. In circumstances where technological progress on the part of large firms has been arrested by a tactical unwillingness on their part to adopt and develop exiting basic scientific discoveries (or invent totally new technical solutions), entrepreneurs, who have no inertia causing interest in the *status quo*, often destroy existing technological paradigms and clear the way for a new period of rapid technological change–led growth (Schumpeter, 1939; Dosi, 1993). These individuals are often disenchanted 'spin off' entrepreneurs from existing large firms (Oakey et al., 1988) or academic entrepreneurs from universities (Roberts, 1991).

The technical and/or managerial fragility of the *status quo* maintained by large firms can often be detected in the simplicity of the technical change that is used by entrepreneurs to destroy current dominant paradigm and create market share for the new entrepreneur entering long-established areas of product- or service-based industry. For example, James Dyson, after many years of sectoral stability, revolutionised the vacuum cleaner industry by inventing the 'bag-less' vacuum cleaner. Similarly, Peter Wood launched 'Direct Line Insurance' in 1985, which grew to dominate the UK insurance sector by using well-established telesales techniques to sell insurance by telephone, an approach that proved highly popular with customers, while delivering operating efficiencies to the insurance provider. In many ways, the simple nature of these inventions and/or innovations, which significantly could have been easily introduced by large industry leader firms, indicate how complacent many large firm dominators of a sector of industry or commerce can become, rendering them an easy target for aggressive new entrepreneurs, who initially operate from a relatively poor competitive capital base.

In other instances, the invention of new technologies can have very radical impacts on existing product- or service-based industries. In some cases, the new radical technology arises from within an existing sector and is known, but not properly exploited, by major players in this industry. For example, the inability of IBM to adequately respond to the potential of microprocessor, of which they already had detailed knowledge, as a basis for desktop computers in the early 1970s (partly because they were distracted by historically high profits gained from the sale of the mainframe computers that desktop computers were destined largely to replace) is an example of this phenomenon. In other instances, entrepreneurs seek to exploit a new technology by *invading* sectors where a new and damaging technology did not previously exist. Recent high technology examples of this tendency are the destruction of traditional photography (camera and print developing technologies) by digital camera and printing technology, while the Internet has impacted on many areas of business such as postal deliveries, high street shopping and the downloading of music (Christensen, 1997; Christensen & Bower, 2004). While the penetration rates of many of these new Internet-based technologies depend, to a large extent, on adoption levels and rates of related enabling technologies (e.g. desktop computers and broadband internet connections), their eventual importance cannot be doubted. Significantly, many of the firms leading this broadly based new technology revolution are entrepreneurially based new enterprises, beginning from modest resource bases, that have grown large at a very fast rate on the basis of specialist skills that impact upon sectors where they are not well understood (e.g. Google, E-bay and Cisco Systems).

Clearly, existing large firms, who have suffered in this conflict with new firm entrepreneurs, do not willingly surrender lucrative areas of business to new entrants. It has previously been mentioned that complacency and a desire to preserve monopolistic or oligopolistic profits derived from a well-established and heavily invested in technology all contribute to a failure to efficiently keep pace with technological progress in their own and relevant related areas. However, particularly in modern technologically complex and rapidly changing circumstances, the process of choosing the correct path for investment in R&D in large firms is inherently difficult and subsequently risky. Moreover, it is a characteristic of many of the most radical technological revolutions that they often emerge on the boundary between technological disciplines or where technologies intermingle and overlap. Here, many small firms exist, while the core large firms of the overlapping two (or more) sectors involved tend to occupy their respective distant 'centre grounds'. Indeed, as noted above, many of the most damaging technological revolutions emerge from totally different technological areas to that of the large firm impacted by such change. This implies that the affected large firm could not have avoided the damage involved by such penetration through traditional *intra-sectoral* R&D, since such research would have not been directed at the correct area of related science. For example, it is possible that future development of computing 'brain' power will take place in biological science through research on how human brains process information biologically and not through the further development of electronic-based solutions, especially since the potential for further efficiency gains in performance through the progressive miniaturisation of electronic circuits are reaching their physical limit.

While corporate venturing allows large existing firms scope for investing in new firms working in adjacent areas of technology of potential future relevance to their main area of operation (e.g. the interest of large pharmaceuticals firms in their small biotechnology counterparts) (Oakey, Faulkner, Cooper, & Walsh, 1990), the relevance of technical entrepreneurs to the development of new areas of technology in young industries, and radical technological departures from well-established areas of production, remains strong. It is the key role of technical entrepreneurs to expose gaps and weaknesses in the *status quo*, invent new and better ways of meeting existing needs and creating new needs that currently do not exist; especially in circumstances when large existing firm cannot, or will not, perform this function. In terms of economic theory, it is the role of the entrepreneur to *repair* the economic system when the competition-driven process of technological progress may be arrested (for the reasons discussed above). Put simply, in this way, entrepreneurs keep the economic system 'honest'.

Conclusions

Technological progress, by its very nature, is famously difficult to foster or predict, in terms of either the pace or direction of change. There is no specific 'secret formula' for reliably creating successful new inventions or producing the next significant 'breakthrough'. However, governments may beneficially influence the *climate* for technological change in a number of general ways that create the best conditions for progress.

First, in circumstances where technological progress has been stalled by a genuine technical problem that is inhibiting scientific progress and human welfare (e.g. a cure for cancer or HIV in medicine), governments should, through proactive invention and innovation 'push' behaviour, be prepared to fund and support speculative R&D in public institutions, large firms and create new high technology small firm enterprises, targeted at these problems. In cases where the risks to the private sector are prohibitively high for the development of 'blue sky' solutions, and where progress is desirable on humanitarian grounds (e.g. in medicine), national and international resources should be deployed to ensure, in keeping with the basic premise of this chapter, that beneficial technical progress is not arrested. In this context, the recent tendency of many western governments to transfer public sector research funding in universities from 'basic' to 'applied' work can be seen as unwise and 'short-termist' in nature.

Second, governments should be alert to the problems of actual or virtual technological monopolies. Market domination, often monitored by anti-trust government bodies, is only a downstream ramification of technological dominance in which producers, who gain an early control of a powerful and pervasive technology, may seek to monopolise technological market areas by dominating the pace and direction of technological change to their own and not their customer's best interests (the case of Microsoft and the provision of incorporated internet search engines within its Windows software might be cited here as an example). Any large firm that attempts, either working alone or oligopolistically, to arrest technological change at either national or international levels, or to dictate it's pace or direction against the public interest, should be strongly penalised by governments. Nationalistic tendencies, which often emerge to protect major transgressor 'flagship' national producers in the short term, will lead to a poorer overall world technological performance in the longer run.

Third, the industrial development arms of governments in developed economies should ensure that highly conducive environments exist for the formation and growth of entrepreneurial new high technology businesses. Although small in resource terms, it is now well established through many research studies that these new small firms regularly produce new 'paradigm destroying' technological improvements that large firms are often reluctant to contemplate or introduce. The provision of necessary logistical support including management training, premises and marketing support are all relevant to the success of this type of firm. However, the most significant resource is patient and reasonably priced capital, mainly because capital has the unique property of transformability into any of the other resources necessary for inventive success (Oakey, 2003). In particular, in high technology small firms, there is a need for the long-term funding of R&D, prior to any market sales, of up to 10 years. While it is rare for such new firms to remain independent and grow to large size in keeping with the 'Apple Computers' model of growth, continuing independence for such catalyst firms is not essential, provided that the technology such small firms introduce is established as a undeniable force within the relevant industrial sector by them before acquisition, and becomes a reality in the market place.

These legislative measures are necessary at both national and international levels to ensure that technology monopolies do not occur. There is a considerable body of evidence to suggest that the development of industrial monopolies are perhaps the biggest flaw of the capitalist system and, without corrective measures by government, will naturally emerge. While it is accepted that monopolies are the enemy of competition, they also can be the enemy of technological progress by blocking the radical inventions and innovations that ensure desirable smoothly disruptive (and consequently incremental) technological change.

References

Cardullo, M. W. (1999). *Technological entrepreneurism*. Baldock, Herts: Research Studies Press Ltd.

Christensen, C. M. (1997). *The innovator's dilemma*. Boston, MA: Harvard Business School Press.

Christensen, C. M., & Bower, J. L. (2004). Customer power, strategic investment, and the failure of leading firms. In: M. L. Tushman & P. Anderson (Eds), *Managing strategic innovation and change*. Oxford: Oxford University Press.

Cyert, R. M., & March, J. G. (1963). *A behavioural theory of the firm*. Englewood Cliffs, NJ: Prentice Hall.

Denison, E. F. (1967) *Why growth rates differ*. Washington DC, Brookings Institute.

Dorfman, R. (1972). *Prices and markets*. New York: Prentice Hall.

Dosi, G. (1993). Technological paradigms and technological trajectories: A suggested interpretation of the determinants and directions of technical change. *Research Policy*, *22*(2), 102–103.

Freeman, C. (1982). *The economics of industrial innovation*. London: Francis Pinter.

Freeman, C. (1986). The role of technical change in national economic development. In: A. Amin & J. B. Goddard (Eds), *Technological change, industrial restructuring and regional development* (pp. 100–115). London: Allen and Unwin.

Geer, D. E. (2003). Monopoly considered harmful. *IEEE Security and Policy*, *3*(3), 14–17.

Grilliches, Z. (1957). Hybrid corn: An exploration of the economics of technological change. *Econometrica*, *25*(4), 501–522.

Harvey, D. (1973). *Explanation in geography*. London: Edward Arnold.

Kamien, M. I., & Swartz, N. L. (1983). *Market structure and innovation*. Cambridge: Cambridge University Press.

Kondratiev, N. (1925). The major economic cycles. *Voprosy Konjunktury*, *1*, 28–79. (English Translation, reprinted in Lloyds Bank Review, No. 29, 1978).

Kuhn, T. S. (1962). *The structure of scientific revolutions*. Chicago: University of Chicago Press.

Magee, B. (1973). *Popper*. London: Fontana/Collins.

Mansfield, E. (1968). *The economics of technological change*. London: Longmans.

Marris, R. (1964). *Managerial capitalism*. London: Macmillan.

Marx, K. (1867). *Capital*. Moscow: Foreign language publishing house, reprinted 1961.

Morse, R. S. (1976). The role of new technology enterprises in the US economy. Report of the Commerce Technical Advisory board to the Secretary of Commerce, January.

Oakey, R. P. (1984). *High technology small firms: Innovation and regional development in Britain and the United States*. London: Frances Pinter.

Oakey, R. P. (1995). *High technology new firms: Variable barriers to growth. London: Paul Chapman Publishing*.

Oakey, R. P. (1985). Agglomeration economies: Their role in the concentration of high technology industries. In: P. Hall (Ed.), *Silicon landscapes* (pp. 94–117). Allen and Unwin.

Oakey, R. P. (1993). Predatory networking: The role of small firms in the development of the British biotechnology industry. *International Small Business Journal*, *11*(4), 9–22.

Oakey, R. P. (2003). Funding innovation and growth in UK new technology-based firms: Some observations on contributions from the public and private sectors. *Venture Capital*, *5*(2), 161–179.

Oakey, R. P., Faulkner, W., Cooper, S. Y., & Walsh, V. (1990). *New firms in the biotechnology industry: Their contribution to innovation and growth*. London: Frances Pinter.
Oakey, R. P., Kipling, M., & Wildgust, S. (2001). Clustering among high technology small firms: The anatomy of the non-broadcast visual communications sector. *Regional Studies*, *35*(5), 401–414.
Oakey, R. P., Rothwell, R., & Cooper, S. Y. (1988). *The management of innovation in high technology small firms*. London: Frances Pinter.
Oakey, R. P., Thwaites, A. T., & Nash, P. A. (1979). The regional distribution of innovative manufacturing establishments in Britain. *Regional Studies*, *14*, 235–253.
Popper, K. (1965). *The logic of scientific discovery*. New York: Harper Torch Books.
Porter, M. (1998). *On competition*. Boston: Harvard Business Review Book.
Pred, A. R. (1965). The concentration of high value added manufacturing. *Economic Geography*, *41*, 108–132.
Riley, R. C. (1973). *Industrial geography*. London: Chatto and Windus.
Roberts, E. B. (1991). *Entrepreneurs in high technology*. Oxford: Oxford University Press.
Rothwell, R., & Zegveld, W. (1981). *Industrial innovation and public policy*. London: Frances Pinter.
Rothwell, R., & Zegveld, W. (1982). *Innovation in the small and medium sized firm*. London: Frances Pinter.
Schmookler, J. (1966). *Invention and economic growth*. Boston: Harvard University Press.
Schumpeter, J. (1939). *Business cycles, a theoretical, historical and statistical analysis of the capitalist process* (2 Vols.). New York: McGraw-Hill.
Schumpeter, J. (1943). *Capitalism, socialism and democracy*. New York: Harper & Row.
Simon, H. A. (1955). The role of expectations in an adaptive behaviouralistic model. In: J. Bowman (Ed.), *Experimentation, uncertainty and business behaviour*. New York: SSRC.
Smith, A. (1776). The Wealth of the Nations. In: E. Cannan (Ed.), *An inquiry into the nature and causes of the wealth of nations* (5th ed.), Original in 1904. London: Methuen.
Solow, R. (1957). Technical change and the aggregate production function. *Review of Economics and Statistics*, *39*, 312–320.
Swann, P., Prevezer, M., & Stout, D. (1998). *The dynamics of industrial clustering: International comparisons in computing and biotechnology*. Oxford: Oxford University Press.
Thwaites, A. T. (1978). Technological change, mobile plants and regional development. *Regional Studies*, *12*, 445–461.

Chapter 11

Who Builds 'Science Cities' and 'Knowledge Parks'?

Paul Benneworth, Gert-Jan Hospers and Peter Timmerman

Introduction

The recent failure to deliver the Lisbon agenda has led to much soul-searching within Europe (cf. The Sapir Group, 2005). This failure has enlarged the gulf between the limited number of successful knowledge regions, and those regions for whom globalisation has brought further anxiety, job losses and economic restructuring. More recent Lisbon-inspired policies have therefore attempted to build linkages between successful 'knowledge islands' and other, outlying and peripheral places, so that all these areas can benefit from concentrations of European knowledge and innovativeness.

Öresund in Denmark and Sweden, the Eindhoven–Leuven–Aachen triangle in the Rhine–Maas Euregio and the South Ostrobothnia virtual university all attempt to provide 'less successful regions' with access to resources for knowledge-based economic development (Maskell & Törnqvist, 1999; Sotarauta & Kosonen, 2004; Hospers, 2005). Alongside this, there has been an increased emphasis on the physical development of the new knowledge economy by creating spaces for these high-technology developments through science cities, knowledge parks and innovation centres across Europe (Hospers, 2005).

This knowledge-based development model begins from the concept that global knowledge flows can be diverted from pools of success into high-technology spaces in less successful regions. Previous waves of science parks have failed to address the core/periphery problem that produces booming totemic sites of the new economy and drab office developments elsewhere (Massey et al., 1992). However, new theories of economic development emphasise the connecting of 'global knowledge flows' with local activities, through a 'local buzz' that can stimulate new and

New Technology Based Firms in the New Millennium, Volume VII
Edited by R. Oakey, A. Groen, G. Cook and P. van der Sijde

innovative high-technology combinations that deliver economic growth (Chapman, MacKinnon, & Cumbers, 2004; Hospers, 2004; Bathelt & Boggs, 2005).

There is a need for increased local innovative capacity to absorb these global resources, and clearly, high-technology small firms (HTSFs) can help create this 'local buzz'. In this chapter, we consider the way that university spin-off companies can mobilise networks and communities that occupy this physical infrastructure, science parks, learning regions and knowledge cities, and capture and locate 'global resources' in these less successful places. Drawing on work undertaken in the older industrial regions of North East of England and Twente in the Netherlands, in this chapter we highlight the diverse ways in which HTSFs can bring such regions into the global knowledge economy. A dynamic model of community building is elucidated and used to reflect on the relationship between HTSFs and knowledge-based growth, to extend the debate concerning the value of inter-regional knowledge sharing to promote economic development.

Global Pipelines and Local Buzz: A Review of the Literature

There is increasing acknowledgement that knowledge is important to the production process. A series of macro-economic studies have demonstrated that productivity growth has become increasingly dependent on investments in intangible forms of capital (i.e., not land, labour or machinery) and that 'knowledge capital' has increasing returns to scale (Romer, 1994; Solow, 1994; Temple, 1998). Increasing returns to scale suggests that knowledge capital investments will be concentrated in places with competitive advantages in knowledge production, and the rise of a limited number of mega-cities has been linked to this phenomenon (Smith, 2003).

Although the implicit regional consequences of the knowledge economy are growing geographical differentiation and competition through innovation, there is some unease that straightforward knowledge capital narratives are unhelpful for understanding 'ordinary places' (Armstrong, 2001; Moulaert & Sekia, 2003). Moulaert and Nussbaum (2005) argue that knowledge capital encompasses resources without 'economic value' and that other types of capital (i.e., human, institutional and ecological) can also promote territorial development.

Such capitals can help to bring new financial investment into such regions, which produce local benefits in what Bathelt, Malmberg, & Maskell (2004) term 'global pipelines and local buzz'. It is not merely the investment that is important but that the investment allows local actors to have some control over its expenditure, and sufficient time is allowed for these benefits to diffuse regionally (Asheim & Herstad, 2005). Cooke and Piccaluga (2004) describe a 'regional knowledge laboratory' as various actors bringing external investments into a region, which create unique assets that are of value to their external partners.

A problem for less successful regions is a lack of globally connected actors able to bring investments to such regions since large firms tend to be disinvesting and downsizing. In this chapter, we focus on universities, important components of regional innovation systems, which are much more uniformly distributed that either firms or government research laboratories. Drawing on this knowledge laboratory concept, less-favoured regions (LFRs) lack strong knowledge exploitation sectors that can convert global academic prestige and research grants into premium products and export income. We consider two cases of universities which have attempted to produce their own knowledge exploitation sectors by promoting university spin-out companies.

Background to the Study and Methodology

In this chapter, we present two case studies characterised by regionally engaged universities with regional development strategies attempting to exploit university capacities. In the North East of England, a partnership of Newcastle City Council, Newcastle University and the Regional Development Agency (RDA) have jointly purchased a central former brewery site for £30m (€45m) on which to develop a new science campus, 'Science Central'. In Twente, in the east of the Netherlands, the RDA, the University of Twente, its host municipality of Enschede, and a number of other regional bodies have announced support for a 120-ha science park adjacent to the campus entitled *Kennispark*. In both cases, national governments have provided moral support without necessarily providing funding for the schemes that are currently under development.

In both cases, regional partnerships are attempting to promote university-based high-technology growth at a scale not previously achieved, and a number of relatively small-scale successes in university commercialisation have occurred. These various elements have been combined discursively by regional political actors (including each university) to argue that success is possible on a far greater scale. In this chapter, we trace these narratives, to explore whether the elements involved are likely to combine together to produce results at the expected scale. We begin by looking at the novel regional capacities produced by commercialisation, particularly in terms of regional networks of HTSFs that have occurred around each university.

The University of Newcastle, North East of England

The North East of England underwent industrialisation from the late 18th century onwards and has latterly experienced industrial decline in which the economy became dominated by mature mass production businesses with little indigenous entrepreneurship. Newcastle University was formally created in 1963 from King's College Durham, itself formed in 1937 from a specialist marine engineering and agriculture college and schools of Medicine and Dentistry (Loebl, 2001). Agriculture, medicine and engineering were all applied subjects, and King's College reflected this disciplinary mix in its ethos as 'a place of useful knowledge' (Potts, 1998). Despite a prevailing isolationist academic norm and successive UK governments discouraging university/regional industrial engagement from the 1940s to the late 1970s, Newcastle-based academics maintained industrial contacts throughout this period (Potts, 1998).

After 1979, the new European Regional Development Fund 'non-quota' (i.e., community-wide) policies were based on mobilising indigenous business assets for innovation. With few private or governmental R&D organisations active in the North East, the Department of Trade and Industry and local authorities demanded that universities should become actively involved in regional engagement (Benneworth, 2002). This initiated a stream of activities as Newcastle University expanded its regional engagement, including a Micro-Electronics Applications Research Institute (MARI, 1983), an City Technology Centre (1984), a seed capital fund (NUVentures, 1987), a regional development office (1995) and finally a Business Development Directorate (2003). By 2004, regional engagement had become central in two key institutional documents, the Business Plan and the Estates Masterplan.

The University of Twente, the Netherlands

The Twente region industrialised in textiles and supporting machinery from 1830 onwards and, after WWII, entered a period of secular decline, which by the 1970s had become a crisis. The Technical Polytechnic of Twente (THT)[1] was created in 1961 to increase technical graduate numbers, promote regional textiles renewal and support the Dutch transformation into an advanced manufacturing economy. However, as the 1970s textiles crisis unfolded, the Government seriously debated whether to close THT and refocus scarce public resources on more successful regions and industries (Groenman, 2001).

Under the leadership of Harry van der Kroonenberg,[2] the university reinvented itself, changing its name to the *University of Twente* (UT) and rebranding itself as 'The Entrepreneurial University' in 1985. In parallel, UT pioneered a series of institutional

---

1. THT is the abbreviation derived from the Dutch name for the institution, Technische Hogeschool Twente. Although literally meaning Technical High School, the *Hogescholen* are now part of the higher education system as universities of professional education alongside the scientific universities. However, despite the name, THT was created as a technical university rather than a university of professional education.

2. The Rector Magnificus position broadly equates with the position in UK universities of vice chancellor; however, the governance arrangements in Dutch universities are somewhat different to UK universities. UK universities are traditionally governed by an academic body such as Senate, which appoints the senior managers drawn primarily from promoted professors. In the Netherlands, universities have a small executive board, which reports to (and is appointed by) a supervisory board of stakeholders, including academic representation, but also the government and the Ministry of Education. The Rector Magnificus is the senior academic representative on the executive board with responsibilities for teaching and research; the other positions will typically not be academics and have responsibilities for finance, estates, regional engagement and internationalisation. In practise, there has been a convergence of these two systems as both Dutch and UK universities come to terms with very similar external pressures.

innovations, including a technology transfer office (1979), an incubator unit (1982), student entrepreneurship schemes[3] (1985), knowledge circles (1990), regional venture funds (1996), an open innovation centre (1997) and a 'technology accelerator' (2003). These resources made UT a central partner of the provincial government and RDA, who now seek to build on this last quarter century of high-technology growth.

Study Methodology

Each regional case study has involved two elements, a review of regional information sources and a set of key respondent interviews. A wide range of documents were reviewed, including historical and contemporary reports about both universities and their regional contexts, along with contemporary policy documents and strategic plans from each university, regional partners and national governments.[4] Seventy-five face-to-face interviews were undertaken by the lead author in the two study regions (32 in Newcastle and 43 in Twente)[5] through a snowball approach (cf. Yin, 1994).[6] These interviews included 16 spin-offs in Newcastle and 24 firms in Twente, with the remainder comprising a mix of university management, academic and commercialisation staff, and key regional stakeholders including regional development agencies, networking organisations and representatives organisations. The study took place within the framework of an ESRC project 'Bringing Cambridge to Consett?'.

High-Technology Spin-offs Mobilising Regional Communities

It is widely acknowledged that high-technology entrepreneurship is heavily dependent on networks (e.g., Groen & Jenniskens, 2003). It is unsurprising, therefore, that spin-off company formation, a quintessentially high-technology form of entrepreneurship, involves assembling and drawing upon a range of networks proximate to the particular spin-off entrepreneur (Benneworth & Charles, 2005).

---

3. The most famous of these, about which a great deal has already been written, is the so-called TOP programme, from the Dutch name, *Tijdelijke Ondernemers Programma* or Temporary Entrepreneurs' Scheme. The scheme is open to anyone with a business plan to exploit technologies and know-how in university research groups; in practise this restricts participation to recent graduates and people working in companies that have research collaborations to the university. This scheme has existed since 1985, although it has been tweaked in response to experience and the changing demands of funders.

4. Those documents directly cited in the chapter are included in the bibliography; a full list of documents reviewed is included in Benneworth (2005).

5. More interviews were undertaken in Twente because I did not have a good understanding of the regional development context in Twente, whereas I had just completed a research project on regional science policy in the North East England, which provided comparable contextual information for the North East.

6. I consulted with academics in each institution with a knowledge of spin-offs to identify a core of interviewees, and then the same was extended outwards approaching people recommended by the initial interviewees.

However, a number of university spin-off companies (USOs) contributed more to these networks post-formation than ante-formation. In this chapter, we identify six regionally articulated networks. These networks involved universities, spin-off companies and regional partners, linked the university to the region.

Providing Direct Support for Academics

The first networks were the connections the spin-off firms provided back to the professors who formed the companies. In Newcastle, the dominant model of entrepreneurship meant that in many cases the professor was still actively involved with the spin-out. Here a particular technology subsidy (worth around €100k) allowed spin-off firms to undertake novel research, and the term and conditions made it least risky for small companies to spend it with an academic collaborator; thus, spin-offs generated third-stream income for their professors and stimulated innovative blue-sky research attractive to science funders.[7] In Twente, although a similar instrument has now been introduced, at the time of the research it was not available. Thus, in the Twente case, the main discovered form of interaction was spin-off science projects.

Other spin-off/academic linkages existed; in both regions spin-off firms employed post-graduates and post-docs. In both institutions, a post-doc would work for a USO but maintained an academic relationship with a professor. In one case, a spin-off entrepreneur found his professor a useful source of graduates; when the professor retired, he mobilised a community of employers who funded the university (approximately €50k annually) to maintain this employee source. In some cases, students worked in spin-offs on student research projects in cases where USOs were exploiting inventions which emerged within such projects. One research group had a practice of contacting key firms (including 'their' USOs) when they found potentially commercialisable things peripheral to their main research. They described this as 'throwing [the ideas] over the fence' to them. In every single USO, there was some active link back to the university.

Mobilising Soft Networks

A second set of activities occurred when USOs helped to develop soft networks whose presence helped other USOs and HTSFs to succeed. UT provided three excellent examples of where USOs had worked.

---

7. Has one academic professor noted 'The company will pose a question to me, as research director, "that represents a 30% loss of productivity over the entire year, what can you do about it?" And the answer has turned out to be very, very interesting … There are two things. Firstly, how can you devise solutions, and I tend to go to the DTI and say if we could devise a method … this would increase productivity by 30%, increase profit, lead to growth, so many more people would be employed so we'll try and do that with them. Then you go to the scientific literature and you ask, is there any scientific or any knowledge or mechanisms for measuring [what controls productivity] … you can then devise a programme of pure research to try and get at the mechanism'.

First, TIMP was a group of HTSFs funded by a local development agency for collaborative innovation projects in medical technology (Klein Woolthuis, 1999). This initiative succeeded, and a number of other projects emerged, resulting in the sector being designated a strategic thematic area in the regional science council (the Innovation Platform).

Second, the Twente Technology Circle was launched by UT itself to help its spin-offs sell to large regional companies. However, it evolved over the years into a networking and mentoring organisation.

Third, the Technology Exchange Cell was a virtual product development laboratory at the university, initially resourced by a large regional firm, but whose rapid prototype capacities were subsequently provided by spin-offs from the university.

By contrast, in the case of Newcastle, there were much fewer academic entrepreneurs involved in stimulating networking activities in the region. Although a biosciences network was established (Bio^{2}NET), the region seemed somewhat behind Twente in terms of establishing support networks. There were key *individuals* who provided access to one-off advice/ guidance, and the university began to systematically engage with these key advice providers. The Alchemists was an organisation created by three retiring business service professionals to try to stimulate entrepreneurs to grow through using professional advice more effectively. Newcastle University approached their chief executive to sit on their Equity Committee, which oversaw the formal technology transfer process for university IP being spun-out. Other key regional entrepreneurs with their own networks were recruited to sit on key university committees including the Advisory Board for the Business School and committees dealing with other aspects of regional engagement.

Stimulating Financial Resources

The third area set of networking where USOs in both regions were active was in stimulating the creation of regional venture funds. Both institutions in the early 1990s had invested in a few companies, and some of those investments had proven successful. On that basis of this experience, universities in both regions had engaged with their Regional Development Agencies to help them create Regional Venture Funds to meet USOs' needs. The creation of these funds was justified by previous successes and an ongoing university commitment to spin-off creation.

Newcastle University created a specific fund in the early 1990s to invest in spin-off firms, but this had largely disappeared without widespread impact (Potts, 1998). However, what was significant was that one pharmaceutical USO resulting from this project had granted the university a share of its equity the company was sold in 2002, and the university received a £6 m (€9m) 'windfall', although the university had failed to invest in a computer security firm which was sold for $14 m (€12m). In 2002 the university produced 6 spin-out companies when the RDA was creating a high-technology regional venture fund. As part of the RDA initiative Newcastle University persuaded the RDA to create a special proof of concept fund, which was tailored to the needs of USOs.

In Twente, by the mid-1990s, the TOP scheme had been functioning effectively for a decade and demonstrated that the university could produce companies that would establish and grow near to the university. The RDA, UT and the nearby Polytechnic together created and capitalised Innofunds, a regional venture capital fund. This raised €11m that was invested in two tranches in around 30 companies, and although some investments failed, it was a reasonable success.

Making Commercialisation Supportable

The fourth area where USOs built a community was in making commercialisation a core university function. Although both universities liked the idea of getting additional so-called third-stream funds, in both instances there was ongoing resistance to spending core university funds on activities such as Technology Transfer Offices to support USO activities. USOs in both regions became important actors in persuading universities that commercialisation was worth serious investment. It was not just that they helped academics win more research funding or that academics were lured by the financial rewards from selling successful businesses, academic entrepreneurs became involved in the core missions themselves and showed that commercialisation could help universities achieve their teaching and research goals.

One key example of this was when USO entrepreneurs became involved in larger infrastructure developments, which supported core university research and levered in external funds. There were a number of examples of these large multi-million Euro projects, which won large subsidies partly on the basis of scientific excellence, and partly on the basis of expertise in commercialising that expertise. USO entrepreneurs were involved in various ways, but critically helped to demonstrate that UT and Newcastle were good at commercialisation. Both governments increasingly emphasised science commercialisation, which helped both universities to win large grants, which have in turn, funded the following core academic research facilities:

- The International Centre for Life (ICfL): a life-sciences campus funded by £50m lottery and a £10m science infrastructure fund awards, that integrated university genetics research (medical and sociological) with hospital genetics services, and commercial genetics companies including USOs.
- The Institute for Nanotechnology Exploitation (INEX, Newcastle): originally funded out of university funds to integrate existing research strands; the initiative subsequently won RDA start-up funding; then a £7m DTI grant, followed by another £10m of infrastructure funding for nanotechnology.
- MESA+ (Twente): since nanotechnology was a field where the university had produced some early spin-offs; a joint academic/commercial laboratory facility was developed for nanotechnology including consultancy activities (now spun-off). In the first ten years of life, it has produced 30 spin-offs and has own small development fund; co-ordinates the Nanoned Science Exploitation Programme (€50m).

Producing New Growth

A fifth area where USOs contributed to new regional networking was that a number of the entrepreneurs involved in USOs subsequently became serial entrepreneurs and founded other companies producing serial growth. In Twente, four of the interviewed entrepreneurs were involving in diversification through setting up a network of businesses and joint ventures within an overall holding company structure. There were around 250 jobs created in this four company development sequence. In Newcastle, a very successful pharmaceutical USO spawned three further companies on the basis of the cash produced from the sale of the original company. One medical spin-off had set up a number of subsidiaries as a means of testing new markets, while minimising risk. The design team of an engineering USO left and set up their own business, and they both grew to employ over 50 people by the time of the research. This suggests that the networks formed from USOs had a degree of vitality and dynamism, and were not confined to just self-employed academics anchored around the university.

Perhaps more interesting is the role of USOs in helping traditional companies in mature sectors to reinvigorate themselves and become more engaged in high-technology sectors. In both life sciences and nanotechnology, there were a number of North Eastern mature chemical companies working with Newcastle USOs as part of successful attempts to move into new markets. One company spun off from its parent and now employs over 60 staff, and there are around 200 employed in science-intensive biotechnology jobs in formerly mature chemicals businesses. In Twente, there were a number of branch plants in the region whose survival within the corporate structure was dependent on maintaining unique knowledge-bases that other parts of the wider corporation could not copy. A number of these branch plants had working relationships with spin-offs as well as the university to try to sustain their unique corporate capacities (cf. Technology Exchange Cell).

Stimulating Regional Technology Policies

The sixth and final area where spin-offs mobilised networks was around emerging regional science and technology policies. Benneworth and Charles (2005) identified that spin-offs have a role to play in working with regional science and innovation policy-makers by improving business support's quality and relevance. In both regions, RDAs showed sensitivity to spin-offs' needs by creating new facilities such as high-technology venture funds. However, there was some evidence that much of this 'needs sensitivity' involved RDAs deciding what spin-offs needed rather than working interactively with them.

There were more interactive approaches in Twente where the RDA Technostartners programme drew very heavily on the experiences and academic knowledge built up through the TOP programme. The TOP concept was diffused and extended into other contexts including a remote rural area and a college of middle professional education. In the North East of England, the RDA allowed regional universities to administer their proof-of-concept fund, acknowledging that their commercialisation expertise was as effective as anything the RDA could assemble.

There were some examples of how USOs did become directly involved with reconfiguring policy in support of spin-offs. In Twente, one USO entrepreneur was appointed to the Regional Innovation Platform, albeit as a successful entrepreneur rather than as a USO representative, while one 1980s spin-off entrepreneur ran a state-funded 'seedcorn' fund to the south of Enschede. In the early 2000s, the RDA encouraged micro-clusters of high-technology businesses to come to them by funding several collaborative partnerships, many involving USOs in leading roles. There was less evidence of involvement by North Eastern USOs in shaping their regional environment.

Towards a Community-Building Model: USOs and a Populated Regional Science Park Concepts

How do these various contributions and communities come together to support the idea that a 'science park' policy is of regional significance? This can be considered as a process of stabilisation over a long-term period and images in accepting that:

- high-technology USOs can demonstrate that high-technology entrepreneurship is a valuable vehicle for regional development,
- universities have further potential that can be exploited and
- the growth trajectory of USOs shows that the 'science park' is an appropriate way to manage such knowledge exploitation.

This is the basis for the community-building model where something experimental, small scale and indeterminate becomes common practise, large scale and precise. This renders external partners willing to support and invest in the 'science park' concept. It is possible to highlight four stages in this translation process, with at each stage, the contributions made by USOs playing a role in the outcome, whether the next stage can be progressed, to and what the impacts of these changes are.

Experiments in Entrepreneurship

The first stage in the model occurs when a university begins with its experiments in entrepreneurship, and the first USOs emerge from the university. This may be as a consequence of the university launching a scheme such as the TOP programme, or announcing, as Newcastle University did around 1990, that entrepreneurship was something that professors should be doing, and consequently provided a service for interested academics. The key determinant in successfully progression to the next stage is whether producing spin-offs appears to have unrealised potential that could further be exploited. The runaway success of TOP and external interest in Newcastle University's spin-offs both suggested that there is further potential for action.

If the scheme is a failure in its own terms, then it is likely that the university will abandon the scheme once funding for the programme expires. However, there is also a scenario where the scheme might succeed in its own terms, by creating businesses, jobs and other outputs, and yet fail to meet core teaching and research missions.

However, if spin-offs are dramatically successful, then it is possible that they will offer the university an opportunity to enhance core teaching and research goals. Universities have seen that USOs have the potential to bring in external resources that can be invested in core scientific infrastructure. At this point, both universities became interested in finding more generic structures to make spinning-off companies more institutionally rooted.

Institutionalising Academic Entrepreneurship

A common strategy for embedding spin-offs more firmly within each institution was creating specific hybrid institutes where spin-offs, research groups and other commercial partners could interact around a shared set of facilities. Such institutions allowed spin-offs to benefit from the presence of academic infrastructure, and academics to benefit from the presence of a commercial-quality infrastructure. Cost-sharing between academics and companies, for example, could be used to underwrite investment in large capital facilities, which increased the scope for work potentially undertaken by particular research groups.

Building such facilities required the meeting of two key goals, first that there were clear academic benefits in investing core resources in such facilities, and second, that there was a steady stream of spin-out companies emerging that would continue to lever additional resources to support ongoing academic research programmes. Again, USOs' previous collaborations with university research groups and their demonstration that university spin-out activities could be successful helped to justify the creation of 'hybrid institutions' such as MESA+ , ICfL, the BTC and INEX.

However, hybrid institutes were not always successful. There were examples of 'institutes' in both institutions that did not succeed, in terms of either their commercial mission or their academic mission. Obviously, such failures tended to be rapidly closed down or 'merged' into more successful organisations. Some institutes were successful in terms of their commercial mission, but did not contribute to core academic missions. These activities have tended to be spun-off or privatised, because of the difficulties that higher-level teaching and research institutions have in managing non-academic types of activity.[8]

8. In the mid-1980s, both universities had received significant government funding to establish micro-electronics consultancy centres to help local SMEs adopt new technologies; both centres grew very rapidly on the basis of local demand, but once the funding expired, in each case the university felt that it was employing people (and incurring risks) that added nothing to core teaching/ research missions, and so those institutes were privatised, and became spin-off companies.

These hybrid institutes that have been successful were those which helped to win external funding for academics. This was often translational R&D type funds, which could be spent on fairly basic research. Such funds were often justified because each university had successfully 'applied' basic research, in spin-off companies.[9] Because both universities had been able to make the institute-type model work on at least two quite different occasions, both universities attempted to rebuild themselves organisationally to improve their capacity to produce a stream of these hybrid institutes, and hence lever in significant external funds which underpinned need their core academic missions.

Building an Entrepreneurial Culture

The next stage of the process was to build an entrepreneurial culture within the university, in the sense of a set of capacities to develop a stream of hybrid institutes. This institutional change involved reorganising research groupings to be more 'marketable' to external partners and forcing academic structures to accept commercial income targets and making looking for hybrid institutes a key part of the central and faculty business planning process. Both universities drew on their capacities and knowledge built up in spinning off companies to achieve a cultural change. Although this appears to be an issue of institutional management, in both cases USOs were involved in helping with this institutional change. Spin-off entrepreneurs were involved in various ways in triggering university cultural change by advising universities, bringing them good commercial opportunities, and helping them to implement new structures.

Both institutions had taken some time to reach this position; much of the pioneering work of Harry van der Kroonenberg at UT was not driven forward by subsequent Rectors, and at Newcastle, previous vice chancellors had built entrepreneurial capacity within their executive offices without trying to coerce other academics to become entrepreneurial. What appeared to make the most progress were the scope and the terms on which the university engaged with an external community, including USO entrepreneurs.

On those occasions where cultural change had been attempted but had not taken root, entrepreneurs and universities had 'spoken at' each other; this latest phase of engagement involved the two parties entering into each others' confidence and working together towards a common goal, becoming an entrepreneurial university. This goal was clearly only for the direct benefit of the university, and so finding entrepreneurs to become trusted university partners was a difficult issue. USOs were

---

9. Newcastle University received a £3m grant from the Department of Trade and Industry for the Nanotechnology Manufacturing Initiative, which was explicitly justified in terms of Newcastle's success in producing spin-outs. Likewise, the position of UT as the centre of the Dutch Nanoned programme was a consequence of both the scientific excellence in MESA+ but also the fact that spin-outs from UT had been very important in the predecessor programme, Microned.

a good source of trusted university partners because of their various personal, social and commercial linkages back into the universities.

Reconfiguring Regional Partners

The final stage of the process was that regional partners acknowledged that each university was entrepreneurial, with considerable untapped potential for further commercial exploitation, and was sufficiently well-managed with experience and capacity a relevant area. In Newcastle, the RDA had proposed and invested heavily in a number of science projects that could not effectively be stabilised into a regional innovation system. Thus, the RDA engaged enthusiastically with the university's own regional science concept, which involved rebuilding the campus as a set of hybrid institutes (Science Central). In Twente, a number of regional partners adopted the *Kennispark* concept in 2003/2004 and implicit endorsement of the concept through the national spatial economic strategy ('*Pieken*') effectively made an extension to the science park an obligatory goal for Twente's Regional Science Council (i.e., the Innovation Platform).

In each case, regional partnerships were mobilised, which agreed in principle to fund large-scale strategic investment projects, *Kennispark* in Twente and Science Central in Newcastle, with tens of millions of euros. The partnership funding idea recognised the regional value of a university in being able to produce a series of hybrid institutes. The regional value of these hybrid institutes was in accelerating the numbers of spin-offs. Spin-offs had regional economic value because of their visible direct and network contributions. USOs were important both symbolically as 'claimed successes' by the university, but also 'trusted partners' (including USO entrepreneurs) were important in helping RDAs to believe in the high-technology 'fantasies' underpinning strategic projects.

Towards a Model for Community Building

In Figure 1, we graphically represent this process as a flow chart. At each stage, USOs provide capacities that the key mobilising actor, in this case the university, draw upon to move to the next stage. If there are not suitable sufficient capacities, then progress is not possible, and the developments that have been achieved either collapse or wither; unsuccessful projects are abandoned, peripheral institutes are privatised, or commercialisation targets withdrawn. We argue that, in the two study regions, what has happened is that three sets of barriers have been overcome, making a sufficiently compelling case for investment in a hybrid regional science park concept.

In Figure 1, the USOs are only explicitly visible in the first phase, where they are the result of a commercialisation project whose success enables a series of subsequent developments. However, looking more closely at the two case studies, it is clear that

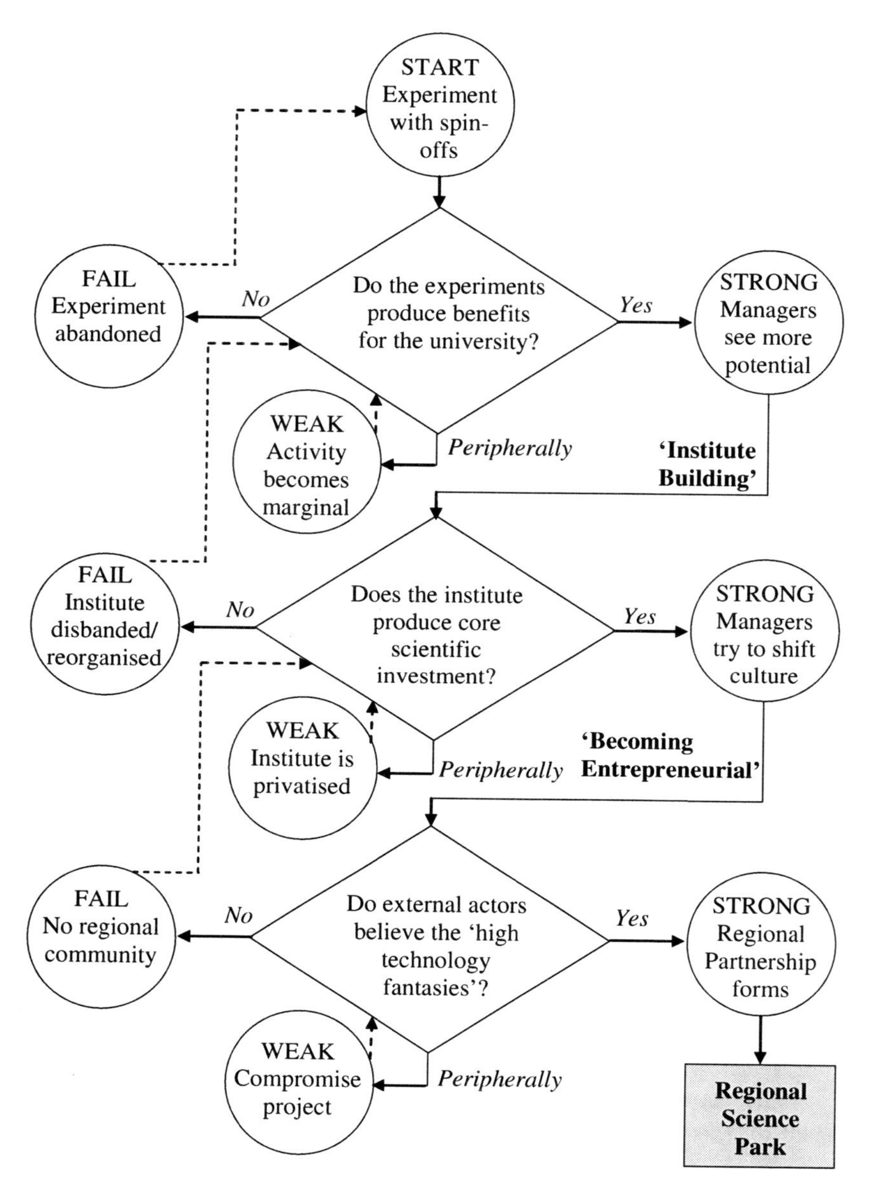

Figure 1: A model of community building with HTSFs and universities.
Source: Authors' own design.

USOs have been involved at *each stage*. USOs have helped universities address the problems at each stage, and to ensure that the institutional trajectory moves from the 'failing' or 'weak' trajectories to the 'strong' trajectory. This implies that USOs are indeed important in the process of progression from vulnerable entrepreneurship projects to a 'regional science park' concept. The contributions at each stage are as follows:

- First, the spin-offs help to make particular entrepreneurship projects succeed; individual entrepreneurs establish businesses, there is mentoring within and between cohorts and results are produced for the projects.
- Second, spin-offs help to make particular hybrid institutions attractive for other investors; on the one hand, they are a rationale for investing to create a critical mass, and on the other hand, they are interesting partners and sources of ideas and employees for other investors.
- Third, spin-offs become involved in helping universities develop the structures necessary for entrepreneurship, including running investment instruments, developing incubator activities and facilitating reorganisation.
- Fourth, spin-offs help to enrol external regional partners by telling them of the value of commercialisation undertaken at the university and its ongoing capacities to exploit its wider knowledge base.

There is therefore a process of co-evolution between the universities and the spin-off communities in the two examples presented. On one side, the spin-offs are evolving from small standalone companies into significant regional actors; and on the other, the universities have evolved from institutions spilling knowledge into their localities into institutions with strategic plans for commercially disseminating their knowledge locally. The result of the coalition is that other local actors create a place where those two activities can take place in parallel, that spin-offs can be actively stimulated, namely the science park.

The consequence of this is that in each case the science park is much more than a piece of real estate. The 'science parks' (Science Central and *Kennispark*, respectively) are a series of relatively stable and certain activities and capacities (spin-out, incubation, mentoring, venture finance, urban regeneration), which are being combined together in an innovative and plausible way. It is by no means certain that the projects will succeed, but the two concepts being developed in the two regions appear to have a considerable advantage with respect to the failed 'high-technology fantasies' of the 1980s, in that they are not beginning *de novo* — the science park in each region is a natural extension of tendencies and capacities already proven and demonstrated.

Of course, these science parks are neither regional knowledge laboratories nor are they fully fledged regional innovation systems. The fact that both universities have worked for twenty years promoting innovation emphasises the difficulties in promoting new regional innovation systems. However, these new science parks are interactive, hybrid spaces where a range of capitals come together and could conceivably provide an arena where local buzz could be created and spread out across the region.

References

Armstrong, P. (2001). Science, enterprise and profit: Ideology in the knowledge-driven economy. *Economy and Society*, *30*(4), 524–552.

Asheim, B., & Herstad, S. J. (2005). Regional innovation systems, varieties of capitalism and non-local relations: Challenges from the globalising economy. In: R. A. Boschma & R. C. Kloosterman (Eds), *Learning from clusters: A critical assessment from an economic-geographical perspective*. Dordrecht: Springer.

Bathelt, H., & Boggs, J. (2005). Continuities, ruptures, and re-bundling of regional development paths: Leipzig's metamorphosis. In: G. Fuchs & P. Shapira (Eds), *Rethinking regional innovation and change: Path dependency or regional breakthrough?* New York: Springer.

Bathelt, H., Malmberg, A., & Maskell, P. (2004). Clusters and knowledge: Local buzz, global pipelines and the process of knowledge creation. *Progress in Human Geography*, *28*, 31–56.

Benneworth, P. S. (2002). *Innovation and economic development in a peripheral industrial region: The case of the North East of England*. Unpublished PhD thesis, University of Newcastle, Newcastle upon Tyne.

Benneworth, P. S. (2005). *Bringing Cambridge to Consett*. Working Paper 2. The University of Twente, Centre for Urban and Regional Development Studies, Newcastle.

Benneworth, P. S., & Charles, D. R. (2005). University spin off companies and the territorial knowledge pool: Building regional innovation competencies? *European Planning Studies*, *13*(4), 537–557.

Chapman, K., MacKinnon, D., & Cumbers, A. (2004). Adjustment or renewal in regional clusters? A study of diversification amongst SMEs in the Aberdeen oil complex. *Transactions of the Institute of British Geographers*, *29*, 382–394.

Cooke, P., & Piccaluga, A. (Eds). (2004). *Regional economies as knowledge laboratories*. Cheltenham: Edward Elgar.

Groen, A., & Jenniskens, I. (2003). *Stimulating high tech entrepreneurship in a region: Many visible hands creating heterogeneous entrepreneurial networks*. Paper presented to 11th High Technology Small Firms Conference, Manchester, England, 12–13 June.

Groenman, B. (2001). Op meisjes was de campus niet toegerust. In: B. Groenman (Ed.), *Van landgoed tot kenniscampus 1961–2001*. Enschede: UT Press.

Hospers, G. J. (2004). *Regional economic change in Europe: A neo-schumpeterian vision*. Münster/London: LIT-Verlag.

Hospers, G. J. (2005). *Slimme streken: Regionale innovatie in Nederland en Europa*. Lelystad: AO Onderwerp.

Klein Woolthuis, R. (1999). *Sleeping with the enemy: Trust dependence and contract in interorganisational relationships*. Enschede, The Netherlands: University of Twente Press.

Loebl, H. (2001). *Outsider in: Memoirs of business and public work in the North East of England 1951–1984*. Gosforth, Newcastle: Fen Drayton Press.

Maskell, P., & Törnqvist, G. (1999). *Building a cross–border learning region: Emerged of the north European Öresund region*. Copenhagen: Handelhøjskolen Forlag.

Massey, D., Quintas, P., & Wield, D. (1992). *High tech fantasies: Science parks in society, science and space*. London: Routledge.

Moulaert, F., & Nussbaum, J. (2005). Beyond the learning region: The dialectics of innovation and cultural in territorial development. In: R. A. Boschma & R. C. Kloosterman (Eds), *Learning from clusters: A critical assessment from an economic-geographical perspective*. Dordrecht: Springer.

Moulaert, F., & Sekia, F. (2003). Territorial innovation models: A critical survey. *Regional Studies*, *37*(3), 289–302.

Potts, G. (1998). *Towards the embedded university?* Unpublished PhD thesis, Centre for Urban and Regional Development Studies, University of Newcastle-upon-Tyne, Newcastle upon Tyne.

Romer, P. M. (1994). The origins of endogenous growth. *Journal of Economic Perspectives*, *8*, 3–22.

Smith, R. G. (2003). World city actor–networks. *Progress in Human Geography*, *27*, 25–44.

Solow, R. (1994). Perspectives on growth theory. *Journal of Economic Perspectives*, *8*, 45–54.

Sotarauta, M., & Kosonen, K. J. (2004). Strategic adaptation to the knowledge economy in less favoured regions: A South Ostrobothnian University network as a case in point. In: P. Cooke & A. Piccaluga (Eds), *Regional economies as knowledge laboratories*. Cheltenham: Edward Elgar.

Temple, J. (1998). The new growth evidence. *Journal of Economic Literature*, *37*(1), 112–156.

The Sapir Group. (2005). An agenda for a growing Europe: The Sapir report. *Regional Studies*, *39*(7), 958–965.

Yin, R. (1994). *Case study research: Design and methods*. Thousand Oaks: Sage.

Chapter 12

Socioeconomic Networks: In Search of Better Support for University Spin-Offs

Danny P. Soetanto and Marina van Geenhuizen

Introduction

Over the past two decades, consideration of the role of universities and other public research institutions regarding the creation of new ventures has considerably increased, among both researchers and policymakers. University spin-offs (USOs) transfer scientific and technological knowledge from universities into the marketplace (Chiesa & Piccaluga, 2000). This contribution may appear small in terms of aggregate employment, but it is a significant contribution to the creation of new jobs and innovation through the diffusion of new knowledge into the regional economy (Rothwell & Zegveld, 1985; Mansfield, 1991). Having new technology as the core of their business and a clustered location, these firms increase the competitive edge of regions (e.g., Keeble & Wilkinson, 1999). In more detail, USOs offer benefits such as the following:

- The promotion of technological entrepreneurship in regions, since they base their businesses mostly on high-technology development and rely on high-technology skills (Shane, 2004).
- Stimulation of other forms of business support and infrastructure that, in turn, provides benefits to other startups (Lockett, Wright, & Franklin, 2003).
- With regard to universities, a strengthening of their relationships with the business community, an improvement of their image, fulfillment of their commitment to society, and income generation from patents (e.g., Heydebreck, Klofsten, & Maier, 2000).

New Technology Based Firms in the New Millennium, Volume VII
Edited by R. Oakey, A. Groen, G. Cook and P. van der Sijde

Over the years, the above advantages have gained wide recognition in such a way that fostering spin-offs has become part of most universities' and research centers' policy. Among the many ways of accelerating the growth of USOs, perhaps the most captivating one is establishing incubator organizations. The networks of incubators have been built gradually in industrialized regions such as the United States and in Western Europe over the past two decades and now have reached maturity (Lalkaka, 2003). The first generation of incubators in the 1980s essentially only offered affordable office facilities to potential new ventures, including shared services, and soft loans. As time progressed, it was realized that the needs of spin-offs included more than just physical and financial support. This situation has challenged some incubators to change into providers of "added-value support," such as business skills training and connecting the entrepreneurs to various networks.

However, after several years of spin-offs enthusiasm, recent studies have started to look critically at the entrepreneurial output of universities. Recent research points out that, even though some successful companies have been created, the spin-off mechanism has been overemphasized, causing remaining doubts over spin-offs survival rates in the long term and benefits to the wider economy (Lambert, 2003). Other studies have also examined such impacts and the high transaction cost of spinning out, in a critical way (Bozeman, 2000; Lerner, 2005). In spite of these critics, the creation of USOs still represents a potentially important innovation mechanism for universities and their regions (Vohora, Wright, & Lockett, 2004), although policy needs to improve to create better performing spin-offs. This means focusing on the quality of spin-offs instead of quantity (Clarysse, Wright, Lockett, Van de Velde, & Vohora, 2005).

In this chapter, we perceive the growth of USOs as a process in which such firms try to acquire vital resources as such resources are often not sufficiently available. Some USOs receive extensive support from incubator organizations while other USOs have to strive to acquire resources on their own. The chance to grow depends critically on the environment of spin-offs and on the nature of their interaction with "external partners" including friends, family, colleagues, former lecturers, and professors who provide access to important resources (Birley, 1985). For this reason, socioeconomic networks should receive considerable attention from incubator organizations in composing support programs. However, so far, little attention has been paid to ways of improving the network structures of USOs. In addition, although networking is seen by various authors as a key feature in explaining the strength of newly established firms, and in predicting their future success (Larson & Starr, 1993), very few studies (except for Perez & Sanchez, 2003) focus on the early years of USOs' network developments. Moreover, with some exceptions (e.g., Nicolaou & Birley, 2003), relatively few empirical studies have attempted to use statistical analysis. Most research on USO networks draws on case studies. Thus, so far, the influence of socioeconomic networks on early growth of spin-offs remains unexplored in terms of medium-sized or large surveys (Markman, Phan, Balkin, & Gianiodis, 2005).

This study has been undertaken in response to the lack of attention to factors underlying the growth of USOs. We address the following question: what is the influence of different socioeconomic networks and different support on the growth of USOs? Do particular network features enhance growth and other ones hamper growth? Following Cooper (1993), we adopt a multi-factoral approach in the exploration of firm performance, instead of focusing only on one particular factor, because such an approach is more comprehensive. Accordingly, we develop a causal model and test this model by applying regression analysis. The chapter has the following structure. First, we provide a brief overview of resource-based theory, followed by the development of hypotheses. In the next section, the research design is discussed. An examination of the results of the regression estimation and an evaluation of these results follow. The chapter ends with a brief discussion of policy implications.

Theoretical Background: Resource-Based Theory

In general, to understand factors that influence the growth of firms, scholars have developed various theoretical viewpoints, including knowledge-based theory, the dynamic capability perspective, and business networks and strategic alliance approaches. Basically, all of them are rooted in the resource-based view (RBV) introduced by Penrose (1959). Penrose's seminal work has been instrumental in the on-going development of modern resource-based theory that is applied in many fields such as strategic management, organization studies, and marketing.

According to the RBV, firms are conceptualized as heterogeneous bundles of assets or resources tied to the firms' management. Firms acquire or search for resources inputs and convert these into products or services for which revenue can be obtained. The RBV suggests that heterogeneity of resources is necessary but not sufficient for a sustainable advantage. Resources should also be unique to create competitive advantage and at the same time difficult to imitate, otherwise, competitors can easily obtain these resources and neutralize such competitive advantage (Barney, 1991).

In reality, young firms frequently lack critical resources. Apart from a lack of *basic resources* such as initial investment and office facilities, they may also lack what are called *added-value resources* such as the ability to identify and explore business opportunities, to learn how to manage the firm, and to achieve business guidance or advice. Such resources can be acquired externally, through networks (Pisano, 1990).

In general, entrepreneurs are part of a network and are dependent on external actors (Pfeffer & Salancik, 1978). Strong relationships with various partners in the network may be an advantage in gaining resources (Hoang & Antoncic, 2003) as they provide entrepreneurs with avenues for negotiation and persuasion and enable them to gather a variety of resources (e.g., market information, problem solving, social support, venture funding, and other financial resources) held by other actors (Nicolaou & Birley, 2003). Birley (1985) observes an extensive use of social networks in the early stages of a venture generation process. Starr and MacMillan (1990) document the use of social and economic exchange mechanisms to acquire resources and to gain legitimacy at the same stages.

In the literature (e.g., Birley, 1985), a distinction is made between two types of networks on which firms can draw; these are formal and informal. Formal networks include financial institutions, accountants, lawyers, the chamber of commerce, small business administrators, and so on. These are the people/institution who are directly connected to business functions. Informal networks may include family, friends, previous colleagues and previous employers, and former professors and lecturers. Contacts with them may range from personal to business issues. Some of the relationships may have been established long before the entrepreneur launched his or her business. However, such networks are not static but dynamic. For instance, informal networks with friends may become formal when friends turn into customers, and formal networks may become informal when business relations mix with personal friendship. Moreover, there is no clear boundary between formal and informal. A person can be a tax consultant and a friend at the same time.

In this study, networks are defined as a socioeconomic network of important "*partners*" that potentially provides valuable knowledge on resources. Such networks refer to personal relations and to business relations with a high personal content and informal way of interaction. As USOs frequently lack critical resources in the early development stage, especially entrepreneurial knowledge and skills (van Geenhuizen & Soetanto, 2004), they may attempt to manage these obstacles by seeking a solution through their "*partners*" (e.g., friends, colleagues, and former professors). Relationships with these "*partners*" may be essential in gathering relevant knowledge to achieve external support and services and to access external resources not available in-house (Birley, 1985). Thus, in the early years of a USOs' development, socioeconomic networks may be important and cannot be neglected. Note that more recently attention in management studies has shifted to the nature of the networks involved, as social benefits may be captured only through particular qualities of the interaction (Hughes, Ireland, & Morgan, 2007).

Another option for gaining resources is through an incubator's support program. As organizations, incubators aim to accelerate the development of startups by providing an array of targeted resources and services. Incubators traditionally merge the concept of fostering new business development with the concept of technology transfer and commercialization (Phillips, 2002). They can be seen as entrepreneurial (non-profit) organizations in performing a bridging function between promising spin-offs and resources required by these spin-offs while protecting them against potential failure (Hackett & Dilts, 2004). In particular, incubators may act as a link between startups and other partners that provide resources such as venture capitalists, government agencies, financial institutions, and other business practitioners. In fact, incubators provide a mechanism for a wide range of networking.

Many incubators employ a central building in which they offer customized rooms and shared services. However, there are also examples of decentralized facilities, spreading over different faculty buildings on a university campus. Generally, incubators support startups only on a temporary basis, for example, 3 or 4 years, after which the startups are forced to leave the incubator and support will end.

Model Development and Hypotheses

The central proposition of resource-based theory is that a firm has to build on and maintain its set of resources to survive and stay competitive. As the resources do not reside exclusively within firms, we argue that there are two ways through which USOs can gain resources, which are (1) through socioeconomic networks and (2) through an incubators' support. The following sections will discuss each of the two factors and will present various hypotheses based on them.

Socioeconomic Networks

Although many empirical studies support the rise of benefits from socio-economic networks on firm growth, little is known about favorable structures and favorable types of relationships in socioeconomic networks. In reality, firms utilize different levels of network resources, while the structure of the networks may vary, as does the social and spatial pattern of the networks. Therefore, the first part of our model deals with the influence of these characteristics on USOs' growth, involving the tightness of the network (structure), the strength of the relationships, the heterogeneity of the social background of the partners, and the spatial proximity of the partners.

Structural Characteristics: Tight or Loose Networks In the literature on small business development, the importance of dense or tight networks is emphasized as being one of the factors influencing the survival of new firms. Tight networks are described as networks in which all partners are connected to each other. If partners know each other well and interact frequently, they are more likely to convey and reinforce norms of exchange and are better able to monitor behavior and enforce sanctions in the network. In business, such networks will reduce risk and enhance the opportunity to build cooperation and get access to resources from other partners connected in the network. Partners in this kind of network are familiar with each other's interests and build trust and credibility in each other. Therefore, tight networks are beneficial for the transfer of complex (fine-tuned) and tacit knowledge, to achieve legitimacy or reputation, and perform joint problem solving (Coleman, 1990; Uzzi, 1996).

In contrast, Granovetter (1992) suggests that partners who are connected in loose networks will enjoy large advantages. Accordingly, a loose network structure causes benefits from diversity of information and brokerage opportunities created by a lack of connection between separate clusters in the networks. This leads into a concept called *Structural hole* (Burt, 1992). Persons who occupy brokerage positions between clusters have better access to information. By being connected in a network, rich in new information and opportunities, entrepreneurs enjoy benefits in terms of (1) enhancing business opportunities, (2) getting access to resources that could not otherwise be obtained, and (3) getting information on partners who may give access to new business networks.

We may conclude that studies on the influence of network structure on the performance of new firms are not conclusive. Several studies stress that linkages with tight networks are more advantageous for the early growth of firms (Gulati, 1995), while others emphasize the importance of being connected to loose networks (McEvily & Zaheer, 1999). This consideration leads to our first hypotheses:

Hypothesis 1a. *The performance of USOs is positively affected by tight networks.*

Hypothesis 1b. *The performance of USOs is positively affected by loose networks.*

Strength of Relationships: Strong or Weak Relationships Whereas the above characteristics refer to the structure of networks, strength refers to the quality of the relationships. The strength of relationships between USOs and their partners varies based on the time invested in the relationships. Usually, strong relationships are based on a long-term and intense interaction. Typical examples of strong relationships include friendship and family ties. However, Granovetter (1995) defines the strength of relationships in terms of time and emotion invested in a relationship as well as reciprocity between the partners. As people know each other better and become emotionally involved, they will develop a relationship in which they put trust, commitment, and willingness to support each other reciprocally. This type of relationship is important for entrepreneurs trying to market an unproven product with limited resources. In such a highly uncertain situation, entrepreneurs will rely heavily on close friends or family members for learning, protection, and support.

However, the concept of social networks presents a contradictory argument. Although initially developed by sociologists, this concept has been increasingly used to explain economic actions (e.g., Larson & Starr, 1993). Granovetter (1973, 1985) argues that new information is obtained through casual acquaintances rather than through strong personal relationships. Since strongly connected partners are likely to interact frequently, much information that circulates is the same. Conversely, *weak ties* often include links with partners who move in social circles other than those of the focal network partners. It is suggested that weak ties are an important source of information about activities, resources, and opportunities in distant parts of the social system. Weak ties are often more important in spreading new information or resources because they tend to serve as a bridge between otherwise disconnected social networks (Elfrink and Hulsink, 2003). In attempts to obtain resources for growth, weak ties may be essential. It is through weak ties that USOs can recognize novel information, which leads them to new resources and enables to exploit new business opportunities. Because of the contradiction in the above arguments, we formulate the following hypotheses:

Hypothesis 2a. *The performance of USOs is positively affected by strong relationships.*

Hypothesis 2b. *The performance of USOs is positively affected by weak relationships.*

Social Characteristics: Heterogeneity of Contact Background The third hypothesis is concerned with the social background of network partners. Marsden (1987) shows that partners from diverse social backgrounds (i.e., integrating several spheres of society) facilitate more beneficial associations than partners from a similar social background. Accordingly, with regard to USOs' development, the more heterogeneous partners USOs have, the larger the variety in resources such as know-how, information, and expertise they can enjoy. Heterogeneity in partners' backgrounds increases the likelihood of obtaining valuable information and knowledge and guides spin-offs to different resources. Therefore, we hypothesize as follows:

Hypothesis 3. *The performance of USOs is positively affected by a large heterogeneity of partners' backgrounds.*

The Spatial Dimension: Geographic Proximity In studies of network creation, it is assumed that networks do not randomly link individuals. Rather, people interact most frequently with those in close geographic proximity and with whom they share common backgrounds, interests, and affiliations (Gertler, 2003). Because both physical and social locations strongly influence people's activities, proximity on these dimensions increases the likelihood of interaction and communication (Blau, 1977). More specifically, close geographic proximity decreases direct costs associated with the frequent and extended interaction necessary for maintaining social relationships (Zipf, 1949), particularly regarding close personal networks. Put it in a slightly different way, a network of partners that is clustered in space provides greater opportunity to actively interact with partners and to benefit from knowledge spillovers compared with a network over large distances (e.g., Audretsch, 1998; Camagni, 1991). As the geographic distance between spin-offs and their partners increases, the opportunity for meeting in person and face-to-face interaction is smaller, and it is more difficult to maintain effective relationships leading to higher costs of coordination. Accordingly, we argue that close spatial proximity between USOs and their partners will positively influence USOs' performance.

Hypothesis 4. *The performance of USOs is positively affected by a close spatial proximity to network partners.*

Type of Support

The nature of support provided to spin-offs may vary considerably, depending on the perceived needs of spin-offs and the competence and resources of incubator organizations. Conventional support is oriented toward the provision of "*basic resources*" (e.g., office, administration, and financial support). However, there has been an important evolution in the kinds of support available, from conventional to added-value support; the latter includes support such as entrepreneurial courses for enhancing the skills of entrepreneurs (such as strategic thinking and negotiation skills), business mentoring, and networking services. Overall, the types of support provided by incubator organizations can be grouped into the following categories:

- Conventional support (e.g., accommodation; loan, grant, venture capital; shared services).
- Added-value support (e.g., business counseling, consultation, entrepreneurial training, networking options, equipment and research facilities).

We assume that conventional support only fulfills the basic needs of USOs. It helps USOs to overcome early obstacles in terms of initial investment and office or laboratory. However, for further development, USOs may need added-value support as well. Therefore, we propose the following hypothesis:

Hypothesis 5. *USOs which receive conventional support plus added-value support perform better than USOs that receive only conventional support.*

Moderating the Impact of Network Building Support

Spin-off companies may receive support by being introduced to socioeconomic networks through the incubator organization or by training to develop own capabilities to establish and maintain network relations. The last capability is increasingly seen as crucial in survival and growth of USOs (Walter, Auer, & Ritter, 2006). Accordingly, we may assume that added-value support, especially receiving support in searching of adequate partners, will moderate the relationship between socioeconomic networks and performance (Figure 1). Therefore, we hypothesize as follows:

Hypothesis 6. *The performance of USOs is positively affected by the interaction effect between socioeconomic networks and network building support.*

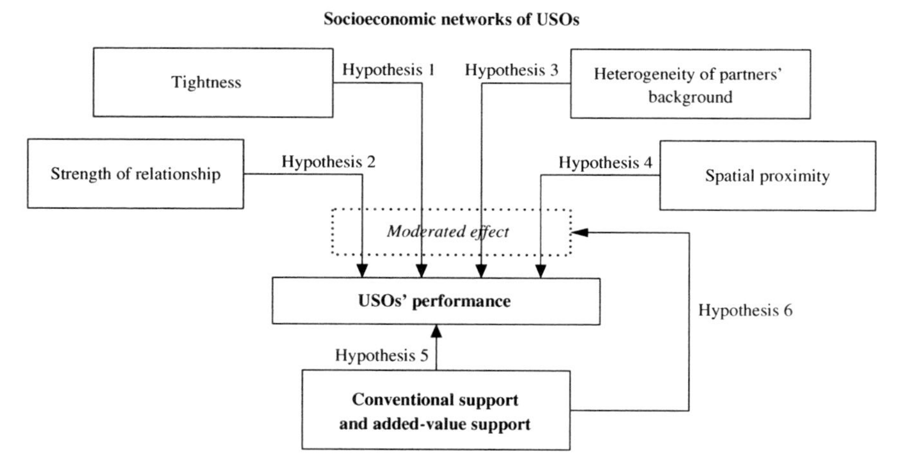

Figure 1: Network and support factors influencing USOs' performance.

Research Design

This study used an interview survey of USOs of Technical University of Delft, the Netherlands (TU Delft). The population included 61 spin-offs, defined by two criteria: that is an age of not more than 10 years and received support, of which at least one support type from TU Delft. The response rate was 67% leading to 41 valid interviews.

In this analysis, Ordinary Least Square (OLS) regression estimation was used (see the Appendix for an explanation of the measurement of the independent variables). To test the hypotheses, four models were estimated. Model 1 tested the hypotheses concerning socioeconomic networks of the USOs. The impact of support was tested in models 2 and 3. Model 2 includes a single variable indicating whether the USOs received merely conventional support or a combination of conventional and added-value support. In the next model (model 3), we examined the influence of each type of support on USOs' performance. Moreover, we also conducted a regression analysis by combining the characteristics of the socioeconomic network and support into one model (model 4).

In a next step, the moderating influence of one single support measure, that is, support in network building, was tested. A moderated hierarchical regression analysis was utilized following the method described by Cohen and Cohen (1983). Accordingly, the change in R^2 was evaluated as an indicator for model significance. In the first step, the characteristics of the socioeconomic networks were entered; in a second step, the support variable was added, and this was followed by adding the interaction effect between support and each of the characteristics of socioeconomic networks.

One of the pitfalls in estimating regression models is the existence of multicollinearity among the independent variables. To check for multicollinearity, the so-called variance inflation factor (VIF) was used, which is the reciprocal of tolerance. Large VIFs are an indication for the presence of multicollinearity. The VIFs ranged from 1.24 to 1.58, meaning that no multicollinearity problems could be identified.

Empirical Results

Characteristics of the Sample

In this section, we first examine the characteristics of TU Delft's spin-offs in the sample, using demographic data and data on innovation and activity (Table 1). With regard to age, the three age categories with borderlines of 4, 6, and 10 years take approximately the same share (i.e., 32–34%). We excluded very young spin-offs from the sample. Nevertheless, TU Delft's spin-offs are relatively small in size, most of them are in the category of less than five full-time equivalent (FTE) employees (44%) and a slightly smaller part is the 5–10 FTE category (37%). Some of the spin-offs can be seen as strongly innovative. Although almost all perform in-house R&D (92%), only a relatively small share invests a large amount in R&D, in that only 27% invests more than 30% annual turnover. The USOs produce turnover from different

Table 1: Characteristics of the sample.

	Frequency (Absolute)	Frequency (Percentage)
Age		
- 2–4 years old	14	34.2
- 4–6 years	14	34.2
- More than 6 years (to 10 years)	13	31.7
Size (fte)		
- Less than 5 fte	18	43.9
- 5–10 fte	15	36.6
- More than 10 fte	8	19.5
In-house R&D	38	92.0
R&D (percent of annual turnover)		
- *Noneorless than 10%*	12	29.3
- 10–30%	18	43.9
- More than 30%	11	26.8
Source of turnover[a]		
- Product sale	22	31.4
- Consultation	24	34.3
- Development and design	24	34.3

[a]More than one category possible per spin-off.

sources, including product sales, consultancy, and development and design. In some cases, spin-offs have more than one source of turnover.

USO companies are usually short of resources. In some cases, this may be so severe that the companies' growth and survival are in danger. Therefore, we will next examine which obstacles to growth have been faced most frequently by the spin-offs in our sample (Table 2).

A shortage of knowledge (skills) tends to be the most problematic need, especially concerning marketing knowledge and sales skills. Dealing with future uncertainty together with a lack of capability in forecasting future markets tend to be among the most frequently encountered obstacles too (at least around 20 spin-offs). Physical obstacles such as a lack of adequate accommodation and lack of research and testing facilities tend to be of minor importance. The previous findings are similar to those of the findings of a study of problematic needs experienced by TU Delft spin-offs in 2002 (van Geenhuizen & Soetanto, 2004).

It is not surprising that shortages in market- and management-related knowledge and skills are the major obstacles, since USOs evolve from an initial idea in a non-commercial environment to a competitive profit-generating firm, a process in which new and completely different knowledge and routines are required (Vohora et al., 2004). To overcome these obstacles, it is plausible that USOs seek external support through their socioeconomic networks.

Table 2: Problematic obstacles.

Obstacles	Specification	Frequency of Occurrence	Share of All Obstacles (%)	Rank
Market-related knowledge	Lack of marketing knowledge	22	15.7	1
	Lack of sales skills	20	14.3	2
	Forecasting of future market	18	12.8	4
Management skills	Dealing with uncertainty	19	13.6	3
	Management overload	12	8.6	5
Market	Lack of market demand	10	7.1	6
Financial	Lack of investment capital	9	6.4	7
	Lack of cash flow	6	4.3	8
Physical	Lack of adequate accommodation	5	3.6	9
	Lack of research/testing facilities	4	2.9	10

Note: More than one obstacle possible per spin-off.

We now move attention to the descriptive statistics of the variables used in our model estimation (Table 3, see also the Appendix). We have measured networks as ego-networks and used the name-generator technique. This technique is conducted by asking the respondent to name individuals with whom they mainly interact or have interacted by exchanging information concerning business problems and opportunities and concerning various types of resources. Consistent with other studies, TU Delft's spin-offs had almost four partners on average.

Renzulli, Aldrich, and Moody (2000) reported that spin-offs have an average of 4.8 external partners. McEvily and Zaheer found 3.5 connected partners per spin-off. According to the respondents, they meet their partners face-to-face on average 2.3 times per month and they had known these partners for 6.0 years on average. With regard to the various types of support, most of the respondents had received support such as training/seminar (63.4%) and loans or grants (51.2%). Almost half of the respondents had been supported with accommodation and shared services (46.3%). Whereas approximately one-third (34–37%) had received at least one of the added-value support types dealing with the market, management, and network building.

Unraveling Factors that Influence Growth

In this section, we examine to what extent the previously discussed socioeconomic networks influence the growth of USOs. In addition, we examine the influence of

Table 3: Descriptive statistics.

	Mean	SD
Dependent variable		
Job growth (fte, 1996–2005)	1.10	1.02
Independent variables		
Characteristics of network		
Number of partners	3.75	0.965
Frequency of interaction with partners	2.28 (per month)	1.26 (per month)
Duration of relationship with partners	6.00 (year)	2.73 (year)
Heterogeneity index	0.58	0.13
Spatial proximity of partners	20.11 (minutes by car)	7.39 (minutes by car)
Type of support	**Absolute**	**Share (%)**
Office and services	19	46.3
Grant, loan, access to venture capital	21	51.2
Market-related support	15	36.6
Managerial consultation	15	36.6
Network building	14	34.2
Training/seminar	26	63.4
Access to research results from university	10	24.4
Access to research facilities at the university	16	39.0

various types of support that USOs have received. To these purposes, regression analysis using four models was performed (Table 4). We start with a discussion of the statistical significance of each beta-coefficient, and this is followed by the interpretation of the sign of each coefficient. Next, we discuss the results from the moderated hierarchical regression analysis. At the end of this section, we draw conclusions concerning the hypotheses.

All coefficients concerning the network characteristics are consistently significant throughout models 1 and 4 and so is the coefficient concerning a combination of conventional and added-value support in models 2 and 4. With regard to the different types of support (model 3), only four coefficients (e.g., concerning office and service, grant/loan or venture capital access, market-related support, and network building) are significant. Model 4 shows, however, that a combination of support is only significant at the lowest level ($p<0.10$). On the basis of a different R^2 of model 1 compared with model 2 (0.72 versus 0.37), we may conclude that network characteristics dominate as influences on the performance of USOs.

Our next discussion will focus on the direction (*sign*) of the regression coefficients. Tight, strong, and proximate networks tend to negatively influence growth, whereas heterogeneous networks tend to influence growth positively. Model 1 reveals that a

Table 4: Regression results (OLS).

Independent Variables	**Model 1**	**Model 2**	**Model 3**	**Model 4**
	β	**β**	**β**	**β**
Socioeconomic networks				
Tightness	−0.32**			−0.30**
Strength of relationship	−0.33*			−0.19*
Heterogeneity in partners' background	0.27**			0.26*
Spatial proximity (of network partners)	−0.44**			−0.41**
Nature of support				
Combination conventional and added value support		0.59**		0.19*
Type of support				
Conventional support				
Office and services			−0.33**	
Grant, loan, or venture capital			−0.30*	
Added value support				
Marketing-related support			0.37**	
Managerial consultation			0.02	
Network building			0.43**	
Training/seminar			0.05	
Access to university research results			0.23	
Access to university research facilities			0.07	
F	22.64	13.89	2.34	20.59
Significance of F ($p<F$)	0.00	0.00	0.04	0.00
R^2	0.72	0.26	0.37	0.75
Adjusted R^2	0.68	0.24	0.21	0.71

*$p<0.05$; **$p<0.01$.

loose network tends to be essential. This implies that having partners in different networks (loose networks) may increase a USOs' performance. With regard to strength of the relationships, employing weak relationships tends to be positive for growth of the spin-offs. Furthermore, there tends to be a positive relationship between USOs' performance and the heterogeneity of contacts. However, the results fail to prove a positive impact of nearby partners on USOs' growth. Apparently, local knowledge spillovers have no positive impact on USOs' growth. Furthermore, with regard to the kind of support, the regression results (model 2) indicate that a combination of conventional and added-value support tends to enhance growth.

The regression model (model 3) reveals different findings concerning the role of individual support measures. Some support such as market-related support and network building are found to positively influence a USOs' performance. The two types of conventional support show a negative sign. Probably, this is caused by the fact that our sample includes a relatively large share of young spin-offs that have received this support but could not grow yet because of their young age (34% falls in the category 2–4 years old). The results of model 4 once more confirm that USOs that receive a combination of conventional and added-value support perform better.

The previous analysis suggests a positive influence of the following characteristics of networks on performance of USOs, that is, loose networks, weak relationships, heterogeneous contacts and a large distance to partners, and of a combination of conventional and added-value support.

A Focus on the Role of Network Support

To explore whether network building support plays a role in moderating the relationships between network characteristics and growth, we employed a two-way moderated hierarchical regression model. The results (Table 5) point that a significant influence only appears with regard to heterogeneity of the partners' social background. Interaction between network support and other network characteristics is not significant. Apparently, socioeconomic networks develop and grow under the influence of other factors, such as the entrepreneur's capability in establishing networks, location history, or relationships already present before the company started.

Summary of Results

We may summarize the above results in terms of rejecting or accepting the hypotheses as summarized in Table 6.

Our study needs to be seen as a first exercise that calls for some more solid testing of the hypothesis investigated here.

Conclusion

The aim of this study was to explore the influence of socioeconomic networks and various support mechanisms on USOs' performance. The empirical findings drawn from a case study of USOs from TU Delft (NL) confirmed that spin-offs facing networks rich in loose and weak relationships tend to be in a better position to grow compared to spin-offs that employ tight and strong relationships. In addition, a large diversity in the social background of partners tends to be essential. The previous results all pointed in the same direction, that is, relatively open information (knowledge) sources and flow have a positive influence on growth of innovative companies. In contrast to our expectation, the findings failed to prove that close

Table 5: Regression results (moderated hierarchical).

Step	1	2	3			
	Socioeconomic Network (B, C, D, and E)[a]	Network Building Support (A)	Interaction Variables			
			A × B	A × C	A × D	A × E
R^2	0.72	0.75	0.75			
ΔR		0.03	0.00			
F		4.24**	0.10			
R^2	0.72	0.75		0.75		
ΔR		0.03		0.00		
F		4.24**		0.16		
R^2	0.72	0.75			0.78	
ΔR		0.03			0.04	
F		4.24**			5.73**	
R^2	0.72	0.75				0.75
ΔR		0.03				0.00
F		4.24**				0.03

Interaction	Coefficient
Network building support × Tightness (A × B)	−0.05
Network building support × Strength of relationship (A × C)	−0.02
Network building support × Heterogeneity in partner's background (A × D)	0.23**
Network building support × Spatial proximity (A × E)	−0.01

**$p < 0.01$.

[a]A, Network building support; B, Tightness; C, Strength of relationship; D, Heterogeneity in partners background; E, Spatial proximity.

spatial proximity of partners has a positive influence on growth. We may understand the result in this context as follows: close spatial proximity may lead to network characteristics that tend to prevent open information and knowledge flow. However, the idea that close proximity apparently does not lead to cost advantages, such as in achieving new knowledge (knowledge spillovers), is more difficult to understand. With regard to support provided by incubator organizations and/or universities, the results showed that conventional support is effective if combined with added-value support. We may conclude that the overall findings confirm the ongoing tendency discussion in the literature concerning the need to improve the support provided to USOs to enhance stronger growth.

We believe that our case study of TU Delft's spin-offs can be generalized to some extent, namely to technical universities and technical faculties at general universities in Western Europe, particularly in countries facing a relatively weak entrepreneurial

Table 6: Summary of hypotheses investigation.

Hypotheses	Description	Conclusion
Hypothesis 1a	Performance is positively affected by tight networks	Rejected
Hypothesis 1b	Performance is positively affected by loose networks	Accepted
Hypothesis 2a	Performance is positively affected by strong relationships	Rejected
Hypothesis 2b	Performance is positively affected by weak relationships	Accepted
Hypothesis 3	Performance is positively affected by a large heterogeneity of partners' backgrounds	Accepted
Hypothesis 4	Performance is positively affected by a close spatial proximity between USOs and their partners	Rejected
Hypothesis 5	Conventional support plus added-value support gives a better performance than merely conventional support	Accepted
Hypothesis 6	Performance is positively affected by the interaction effect between network building support and socioeconomic network characteristics	Mostly rejected

(risk-avoiding) culture. However, there are two features that make this case study more specific. First, from the perspective of the type of incubation policy, TU Delft's incubation policy represents the so-called low selective model (Clarysse et al., 2005). This model aims to create as many startups as possible and focuses on providing conventional support (although more recently access criteria to the incubation program have become more rigid and added-value support is now quite common). Secondly, from a location perspective, TU Delft's spin-offs are close to two important large cities, The Hague and Rotterdam. This polycentric pattern of cities may lead to different spatial networks compared with spin-offs located near or in single cities.

Despite the interesting results, we acknowledge that there are some limitations in our study. The limitations are methodological in nature. First, we were not able to exclude certain fuzziness in the data on personal networks of spin-offs managed by more than one entrepreneur. This is a common issue in network study using a so-called ego-centric technique (Brewer & Webster, 1999). In conclusion, this study is to be viewed as a first effort to identify broad characteristics of socioeconomic networks of USOs and their influence on growth. Concerning next steps, using an extended sample, the study can test the hypotheses more rigorously, particularly the influence of spatial proximity of network partners on growth of USOs. Moreover, the current study was limited to external factors (networks and support), and in future work the relevant internal characteristics, such as age and size, and strategies, given their potentially moderating effect on socioeconomic networks, need to be identified.

The preliminary outcomes of the current study point to the following measures to improve support for USOs: to focus stronger attention to socioeconomic networks compared with other types of support; to guide network formation and to offer training to increase an entrepreneurs' capability of building and maintaining open networks, exhibiting loose ties, weak relationships, and diversity in social background of partners. There should also be a stronger focus on added-value support to improve the entrepreneurs' capabilities and skills, particularly concerning marketing and sales, and strategic thinking on future uncertainty concerning markets.

Acknowledgement

This study is the result of the Delft Center of Sustainable Urban Areas of Delft University of Technology.

References

Audretsch, D. B. (1998). Agglomeration and the location of innovative activity. *Oxford Review of Economic Policy*, *14*, 18–29.

Barney, J. (1991). Firm resources and sustained competitive advantage. *Journal of Management*, *17*(1), 90–120.

Birley, S. (1985). The role of networks in the entrepreneurial process. *Journal of Business Venturing*, *11*, 107–117.

Blau, M. (1977). *Inequality and heterogeneity: A primitive theory of social structure*. New York: Free Press.

Bozeman, B. (2000). Technology transfer and public policy: A review of research and theory. *Research Policy*, *29*, 627–655.

Brewer, D. D., & Webster, C. M. (1999). Forgetting of friends and its effects on measuring friendship networks. *Social Networks*, *21*, 361–373.

Burt, R. (1992). *Structural holes*. Cambridge, MA: Harvard University Press.

Camagni, R. (1991). Local milieu, uncertainty and innovation networks: Towards a dynamic theory of economic space. In: R. Camagni (Ed.), *Innovation networks: Spatial perspectives* (pp. 121–144). London: Belhaven Press.

Chiesa, V., & Piccaluga, A. (2000). Exploitation and diffusion of public research: A case of academic spin-offs companies in Italy. *R&D Management*, *30*(4), 329–340.

Clarysse, B., Wright, M., Lockett, A., Van de Velde, E., & Vohora, A. (2005). Spinning out new ventures: A typology of incubation strategies from European research institutions. *Journal of Business Venturing*, *20*(2), 183–216.

Cohen, J., & Cohen, P. (1983). *Applied multiple regression/correlation analysis for the behavioral sciences* (2nd ed.). Hillsdale, NJ: Erlbaum.

Coleman, J. S. (1990). *Foundations of social theories*. Cambridge, MA: Harvard University Press.

Cooper, A. C. (1993). Challenges in predicting new firm performance. *Journal of Business Venturing*, *8*(3), 241–253.

Elfrink, T., & Hulsink, W. (2003). Network in entrepreneurship: The case of high-technology firms. *Small Business Economics*, *21*, 409–422.

Gertler, M. S. (2003). Tacit knowledge and the economic geography of context. *Journal of Economic Geography*, *3*, 75–99.

Granovetter, M. (1973). Strength of weak ties. *American Journal of Sociology*, *78*(6), 1360–1380.

Granovetter, M. (1985). Economic action and social structure: The problem of embeddedness. *The American Journal of Sociology*, *91*(3), 481–510.

Granovetter, M. (1992). Economic institutions as social constructions — A framework for analysis. *Acta Sociologica*, *35*(1), 3–11.

Granovetter, M. (1995). Coase revisited: Business groups in the modern economy. *Industrial and Corporate Change*, *4*(1), 93–130.

Gulati, R. (1995). Social structural and alliance formation patterns: A longitudinal analysis. *Administrative Science Quarterly*, *40*, 619–652.

Hackett, S. M., & Dilts, D. M. (2004). Systematic review of business incubation research. *Journal of Technology Transfer*, *29*(1), 55–82.

Heydebreck, P., Klofsten, M., & Maier, J. C. (2000). Innovation support for new technology based firms: The Swedish technopol approach. *R&D Management*, *30*(1), 89–100.

Hoang, H., & Antoncic, B. (2003). Network based research in entrepreneurship. A critical review. *Journal of Business Venturing*, *18*, 165–187.

Hughes, M., Ireland, R. D., & Morgan, R. E. (2007). Stimulating dynamic value: Social capital and business incubation as a pathway to competitive success. *Long Range Planning*, *40*, 154–177.

Keeble, D., & Wilkinson, F. (1999). Collective learning in regionally clustered high-technology SMEs in Europe. *Regional Studies*, *33*, 295–303.

Lalkaka, R. (2003). Business incubators in developing countries: Characteristics and performance. *International Journal of Entrepreneurship and Innovation Management*, *31*(2), 31–55.

Lambert, R. (2003). *Lambert review of business-university collaboration*. London: HM Treasury.

Larson, A., & Starr, J. (1993). A network model of organization formation. *Entrepreneurship Theory and Practice*, *17*(1), 5–15.

Lerner, J. (2005). The university and the start-up: Lessons from the past two decades. *Journal of Technology Transfer*, *30*, 49–56.

Lockett, A., Wright, M., & Franklin, S. (2003). Technology transfer and universities' spin-out strategy. *Small Business Economics*, *20*(2), 185–201.

Mansfield, E. (1991). Academic research and industrial innovation: An update of empirical findings. *Research Policy*, *26*, 773–776.

Markman, G. D., Phan, P. H., Balkin, D. B., & Gianiodis, P. (2005). Entrepreneurship and university-based technology transfer. *Journal of Business Venturing*, *20*(2), 241–263.

Marsden, P. V. (1987). Core discussion networks of Americans. *American Sociological Review*, *52*, 122–131.

McEvily, B., & Zaheer, A. (1999). Bridging ties: A source of firm heterogeneity in competitive capabilities. *Strategic Management Journal*, *20*, 1133–1156.

Nicolaou, N., & Birley, S. (2003). Social networks in organizational emergence: The university spinout phenomenon. *Management Science*, *49*(12), 1702–1725.

Penrose, E. (1959). *The theory of growth of the firm*. Oxford: Blackwell.

Perez, M., & Sanchez, A. M. (2003). The development of university spin-offs: Early dynamics of technology transfer and networking. *Technovation, 23*(10), 823–831.

Pfeffer, J., & Salancik, J. (1978). *The external control of organizations*. New York: Harper & Row.

Phillips, R. G. (2002). Technology business incubators: How effective as technology transfer mechanisms? *Technology in Society*, *24*(3), 299–316.

Pisano, G. P. (1990). The R&D boundaries of the firm: An empirical analysis. *Administrative Science Quarterly*, *35*, 153–176.

Renzulli, L. A., Aldrich, H. E., & Moody, J. (2000). Family matters: Gender, networks, and entrepreneurial outcomes. *Social Forces*, *79*(2), 523–546.

Rothwell, R., & Zegveld, W. (1985). *Reindustrialization and technology*. Harlow, UK: Longman.

Shane, S. (2004). *Academic entrepreneurship: University spinoffs and wealth creation*. Cheltenham: Edward Elgar.

Starr, A. S., & MacMillan, I. C. (1990). Resource co-option via social contracting resource acquisition strategies for new resources. *Strategic Management Journal*, *11*, 79–92.

Uzzi, B. (1996). The sources of consequences of embeddedness for the economic performance of organizations: The network effects. *American Sociological Review*, *61*, 674–698.

van Geenhuizen, M., & Soetanto, D. P. (2004). Academic knowledge and fostering entrepreneurship: An evolutionary perspective. In: H. L. F. Groot, P. Nijkamp & R. R. Stough (Eds), *Entrepreneurship and regional economic development*. Massachusetts: Edward Elgar.

Vohora, A., Wright, M., & Lockett, A. (2004). Critical junctures in the development of university high-tech spinout companies. *Research Policy*, *33*, 147–175.

Walter, A., Auer, M., & Ritter, D. (2006). The impact of network capabilities and entrepreneurial orientation on university spin-off performance. *Journal of Business Venturing*, *21*(4), 541–567.

Zipf, G. K. (1949). *Human behavior and the principle of least-effort*. Cambridge, MA: Addison-Wesley.

Appendix

Regression	Variables	Description
Linear regression and moderated hierarchical regression	Job growth	Measured by average annual growth in full-time equivalents
	Tightness	Measured by the number of existing relationships divided by the number of potential relationships. The value of this variable is between 0 and 1 A low value indicates a loose network and a high value indicates a tight network
	Strength of relationship	Measured by two criteria: the average frequency of interaction and the number of years the relationships have lasted = (no. of frequency of interaction/no. of partners) + (no. of number of years the relationship/ no. of partners) A low value indicates a weak relationship and a high value indicates a strong relationship
	Heterogeneity	Measured by the sum of the outcomes of the heterogeneity index of each type of partners' background (e.g., academic, business). The index is calculated on the basis of the square of the number of partners from a similar background divided by the total number of partners. 1 - (the square of the number of partners from a similar background divided by the total number of partners). A high value indicates that many partners have a diverse background and a low value indicates that many partners have a similar background.
	Spatial proximity	Measured by the average travel time to partners = (no. of travel time/no. of partners) A low value indicates a close proximity and a high value indicates a large distance
	Network building support	The variable is a dummy variable for the network building support (1 = received support; 0 = did not receive support)
Linear regression	Mixed support	The variable is a rank variable for the types of support. The codification is as follows: Rank 1: Only conventional support Rank 2: Conventional and 20% added-value support or only 60% and more added-value support Rank 3: Conventional and 40–60% added-value support Rank 4: Conventional and 60% and more added-value support
	Support types (eight)	Each variable is a dummy variable (1 = received support; 0 = did not receive support)

Chapter 13

A Conceptual Framework for Studying a Technology Transfer from Academia to New Firms

Igor Prodan, Mateja Drnovsek and Jan Ulijn

Introduction

Global technological competition has made technology transfer from academia to firms an important public policy issue (Rahm, 1994). Academia and individual academic institutions are a primary source of new knowledge production and innovation (Brennan & McGowan, 2007). It is widely acknowledged that the commercialization of scientific and technological knowledge produced in public funded research institutions, including universities and research centres, into the marketplace have a fundamental role to play in wealth creation, supporting economic growth and technological innovation, and plays a significant role in new venture creation, growth of existing firms, and new job creation (Mansfield, 1991; Harmon et al., 1997; Ndonzuau, Pirnay, & Surlemont, 2002; Siegel, Waldman, Atwater, & Link, 2003b; Steffensen, Rogers, & Speakman, 1999; Walter, Auer, & Ritter, 2006; Perez & Sanchez, 2003). Research by Acs, Audretsch, and Feldman (1992), Jaffe (1989), Mansfield (1991, 1998), and others indicates that technological change in important segments of the economy has been significantly based on knowledge that spin-off from academic research.

Academic spin-offs are an important means of technology transfer from an academic organization (Gregorio & Shane, 2003; Roberts & Malone, 1996; Nicolaou & Birley, 2003a), a key mechanism for maintaining industry science links (Debackere & Veugelers, 2005), an important means of regional economic development (Mian, 1997; Nicolaou & Birley, 2003a) and are an important mechanism for introducing new commercial products to the marketplace (Association of University Technology

New Technology Based Firms in the New Millennium, Volume VII
Edited by R. Oakey, A. Groen, G. Cook and P. van der Sijde

Managers, 2002). For example, spin-offs are the main mechanism for the rapid growth of high-technology clusters such as Silicon Valley, in San Francisco and Route 128 in Boston (Rogers & Takegami Shiro, 2001). Carayannis, Rogers, Kurihara, and Allbritton (1998) quote a Bank of Boston survey (BankBoston, 1997), which observed that the Massachusetts Institute of Technology had spin-off approximately 4,000 companies, employing 1.1 million people and generated annual worldwide sale of 232 billion US dollars. Furthermore, Mustar (1997) reported that 200 academic spin-off from France that he has studied created 3500 jobs. Spin-offs are also longer term mechanism for the creation of emerging new industries in the long run (Roberts, 1991).

Although many start-ups may fail within a few years as the technology itself fails to prove viable and financiers pull out, on occasion university-based start-ups may grow into major industrial contenders. Several major employers in the San Francisco Bay area that were university generated include Sun Microsystems, Cisco Systems, Chiron, and Genentec. Most often, the results of high-technology start-ups are moderate when successful start-ups based on university technologies are acquired and absorbed by larger companies which acquire the technology or expertise developed by a start-up firm to complement their own R&D initiatives (Graft, Heiman, & Zilberman, 2002).

Politicians in the European Union have also recognized the importance of technology transfer from academia and establishment of spin-offs. Therefore, the European Union funds projects such as PROTON (pan-European network of Technology Transfer Offices and companies affiliated to universities and other Public Research Organizations), PRIME (Policies for Research and Innovation in the Move toward the European Research Area), and INDICOM (Direct indicators for commercialization of research and technology) that examine issues concerning technology transfer from academia, and the establishment of academic "spin-offs" (Lockett, Siegel, Wright, & Ensley, 2005).

Policymakers in many developed countries have also responded to the importance of academic spin-offs by creating infrastructures intended to facilitate the commercialization of scientific research output (Goldfarb & Henrekson, 2003). For example, to stimulate the commercialization of university-based research and promote spin-offs, the UK government established the £50 million "University Challenge," which provides venture capital funding for university-based spin-offs and through a project entitled the Science Enterprise Challenge created 12 Government sponsored Science Enterprise Centres at UK universities, which provide educational, training, and financial services to would-be academic and graduate entrepreneurs (Wright, Birley, & Mosey, 2004; Lockett et al., 2005).

The aims of this theoretical chapter are two fold: (a) to provide a conceptualization of the constructs necessary for studying technology transfer processes and (b) to develop an empirically testable model for technology transfer from academia into new firms.

Technology Transfer Process Conceptualizations

To avoid confusion resulting from various definitions of technology transfer and academic spin-off companies found in previous literature, it is necessary to know how we define these terms in our research.

Technology Transfer

There is no widely accepted definition of technology transfer, but, generally speaking, technology transfer is the sharing of technology, techniques, or knowledge (Melkers, Bulger, & Bozeman, 1993; cited in Phillips, 2002). This also includes sharing know-how and organizational rationalities, which are the "soft" dimensions of technology (Storper, 1995), among individuals, industries, universities, public research institutions, federal, state and local governments, and third party intermediaries. Walsh and Kirchhoff (2002) argue that some form of technology transfer occurs in all organizations among and between departments in the organization, between manufactures and vendors, and between manufacturers and their customers. They additionally state that any technology transfer model developed under competition (which is based on the need for secrecy) must include a high degree of interaction and communication as a necessary ingredient. This is especially true for disruptive technologies since the high level of uncertainty attached to new-to-the-world technologies requires extensive testing. There are other more narrow definitions of technology transfer. For example, Phillips (2002) for the purpose of research on high-technology business incubators, defined technology transfer as the licensing of technology from a university to an incubator client firm, whereas Powers and McDougall (2005) defined university technology transfer as a process of transforming university research into marketable products.

In this chapter, *technology transfer* is defined as the transfer of knowledge, scientific or technical know-how, technology, technology-based ideas or research results, developed within an academic institution, from academic institution to industry, where an academic institution may or may not have the property rights for commercialization of such scientific or technical know-how, technology or research results.

Spin-Off

There have been a number of studies of academic spin-offs worldwide and various definitions are applicable. In one of the first studies on spin-offs, Cooper (1971) used a term spin-off for a new company independent from parent organization, which is often started by a founder, or founders, from the same parent company. According to Cooper (1971), a spin-off company is technological based and emphasizes research and development or places major emphasize on exploiting new technical knowledge.

Garvin (1983) proposed that spin-offs are new firms created by individuals breaking off from existing ones to create companies of their own. A spin-off normally occurs when a firm is formed by individuals leaving an existing firm in the same industry.

Smilor, Gibson, and Dietrich (1990) defined university spin-offs as companies founded by faculty members, staff members, or students who left the university to start a company, or who started the company while still affiliated with the university; and/or based on technology or technology-based idea developed within the university. Similar to this definition is a definition of Steffensen et al. (1999) where a spin-off is a new company that is formed by individuals who were former employees of a parent organization, around a core technology that is transferred from the parent organization. Another similar but different definition of a spin-off is provided by Nicolaou and Birley (2003a), which defines university spin-offs as companies, which involve the transfer of a core technology from an academic institution into a new company where the founding member(s) may include the inventor academic(s) who may or may not be currently affiliated with the academic institution. They additionally explicitly exclude companies established by current or former members of a university, which do not involve the commercialization of intellectual property arising from academic research.

Carayannis et al. (1998) first defined a spin-off as a new company formed by individuals who were former employees of a parent organization, around a core technology that originated at a parent organization and that was transferred to the new company. In the conclusions of their research, they suggested that it is an oversimplification to define a spin-off as a new company in which both the founder and the core technology are transferred from a parent organization, since only one or the other or both of these factors may be transferred. Walter et al. (2006) defines an academic spin-off as a business ventures that is founded by one or more academics who choose to work in the private sector (at least part-time) and that transfer a core technology from their parent organization. Weatherston (1995) described the academic started venture or spin-off as a business venture that was initiated, or became commercially active, with the academic entrepreneur playing a key role in any or all of the planning, initial establishment, or subsequent management phases. Rappert, Webster, and Charles (1999) in their research on academic–industrial relations and intellectual property defined university spin-offs as companies whose products or services develop out of technology-based ideas or scientific/technical know-how generated in a university setting by a member of faculty, staff or student who founded (or co-founded with others) the firm. The individual or individuals may either leave the university to start a company or start the company while still inside the university. It does not matter whether someone was a student or full-time academic and the time interval between the initial research and commercial exploitation is not an issue so long as their university research experience was essential in enabling the firm to provide particular products or services (rather than, for instance, university experience merely providing background knowledge).

Pirnay, Surlemont, and Frederic (2003), based on a literature review, proposed a definition of a university spin-off as a new firm created commercially to exploit some knowledge, technology or research results developed within a university. Similarity, but narrowly, Druilhe and Garnsey (2004) defined the spin-off as a new firm commercializing a proprietary leading-edge technology from a university department and backed by venture capital. Grandi and Grimaldi (2005) proposed a generic definition of university spin-off, which includes cases in which university dependents (academic founders) start a company on the basis of either a university-assigned technology (license on a patented technology) or a more generic area of technological knowledge (non-university assigned). They proposed that a university spin-off also encompasses situations in which the university elects to provide the rights to the technology to an external, independent entrepreneur, non-university-dependent (non-academic) founder, who initiates a new company. Lockett and Wright (2005) narrowly defined university spin-offs as new ventures that are dependent on licensing or assignment of the institution's intellectual property for initiation.

To avoid confusion resulting from various definitions of academic spin-off companies found in previous literature, it is necessary to define what we mean by an academic spin-off company in this research. We define an academic spin-off as a company that is founded (or co-founded by non-academics) by one or more academics (not including students) and was created commercially to exploit some knowledge, scientific or technical know-how, technology, technology-based ideas or research results developed within an academic institution. Such a spin-off would occur where an academic institution may or may not have the property rights for commercialization of such scientific or technical know-how, technology or research results and where it is not necessary that such knowledge, scientific or technical know-how, technology, technology-based ideas or research results developed within an academic institution is a core research focus of an academic institution.

Technology Transfer from Academia to Industry

It is clear that there is more than one mechanism involved in the commercialization of academic intellectual property. Key mechanisms are the formation of spin-off companies, patents, licenses, and research join-ventures (Lockett et al., 2005). Since Jensen and Thursby (2001) found that only 12% of university inventions were ready for commercial use at a time of licence (which points to the importance of incubation (Clarysse, Wright, Lockett, Van de Velde, & Vohora, 2005)) and manufacturing feasibility was known for only 8% and similarly Jensen, Thursby, and Thursby (2003) reported, that faculty involvement in further development is necessary for commercial success for 71% of the inventions licensed, we can be sure that whatever the route of technology transfer is, core to its success will be the role played by the creator of the intellectual property, the individual scientist or engineer (Jensen & Thursby, 2001; Jensen et al., 2003; Goldfarb & Henrekson, 2003; Wright et al., 2004, Markman, Gianiodis, Phan, & Balkin, 2005b). Also, although an innovation may seem clearly marketable, it is often the case that no existing firm will risk taking it on (Graft et al., 2002). Thus, academic spin-offs are important mechanism for transferring technology from academia since the scientist is actively involved in its creation, and validation. In addition, spin-offs based on university technology (such as Lycos, Genentech, Cirrus Logic) tend to survive and are likely to achieve Initial Public Offering (IPO) status (Shane & Stuart, 2002).

Where as in the past academic institutions have passively licensed their technologies to large established companies (Siegel, Waldman, & Link, 2003a), today many academic institution actively search for ways to develop proprietary technology to maximize rents and to spawn new companies (Thursby, Jensen, & Thursby, 2001; Wright et al., 2004; Chapple, Lockett, Siegel, & Wright, 2005; Powers & McDougall, 2005). Licensing, which is the most common mechanism to commercialize university technology (Radosevich, 1995), has the advantage that the academic and the university are able to capitalize on the technology, and the academic is able to pursue his/her research without having to commit large amount of time to commercial matters (Lockett & Wright, 2005). The downsides to this approach are (Franklin, Wright, & Lockett, 2001):

- the nature of the new technology may not be easily patented and transacted through a license agreement and
- universities may not be able to capture the full value of their technology through a licensing agreement and therefore may seek more direct involvement in the commercialization of new technology through spin-off companies.

Despite the perceived importance of spin-offs and growth in the number of spin-offs from universities, there have been very few systematic studies that have examined this phenomenon. In fact, in most research, spin-offs have been one of a number of technology transfer mechanisms under study, including patenting and licensing with relatively little emphasis placed on detailed research into spin-off activity *per se* (Leitch & Harrison, 2005).

A Conceptual Framework for Studying a Technology Transfer Process

Academic entrepreneurship arises from internal as well as external impetuses (Etzkowitz, 2003). Both micro and macro level factors influence the decision to create a new company to exploit an academic invention. At the micro level, research has shown that the motivation of academics' (Roberts, 1991; Steffensen et al., 1999; Shane, 2004), the attributes of technological inventions themselves (Shane, 2001a), inventors' career experiences (Levin & Stephan, 1991), their psychological make-up (Roberts, 1991), and their research skills (Zucker, Darby, & Brewer, 1998) influence this decision. At the macro level, research has shown that technology regimes and characteristics of parent organization (Shane, 2001b; Rogers & Takegami Shiro, 2001; Powers & McDougall, 2005), size of technology transfer office (O'Shea, Allen, Chevalier, & Roche, 2005), age of technology transfer office (Roberts & Malone, 1996; Powers & McDougall, 2005), size of federal funding in science and engineering (Shane, 2004; Powers & McDougall, 2005), level of industry R&D funding (Powers & McDougall, 2005), availability of venture capital (Druilhe & Garnsey, 2004; Powers & McDougall, 2005), the strength of patent protection in a line of business (Shane, 2002), spin-off/parent conflict (Steffensen et al., 1999), the university rewards system, which is based mainly on publications and citations (Goldfarb & Henrekson, 2003; Franklin et al., 2001), university quality (O'Shea et al., 2005), universities' intellectual property (Goldfarb, Henrekson, & Rosenberg, 2001), official university policy toward spin-offs (Chiesa & Piccaluga, 1998; Roberts & Malone, 1996), and government policies (Liu & Jiang, 2001; Shane, 2004) all may influence this decision.

Although both micro and macro level factors influence the tendency of academics to start a new company to exploit academic inventions, we discuss only factors from the entrepreneur's standpoint in this chapter. In developing a conceptual model of technology transfer from academia to new firms (Figure 1), we have included the key facilitators as well as the key barriers to technology transfer previously identified in the literature (mainly literature about general entrepreneurship, about psychology of entrepreneurs, technology transfer, academic entrepreneurship, and university-industry links) and also some additional facilitators as well as barriers to technology transfer, we have identified during previous research. These factors that influence Academic's entrepreneurial involvement or the Academic's intention to become an entrepreneur are personal networks, number of years spent at the academic institution, nature of the research, motivational factors, previous work with the industry, scientific publications, role models, support from academic institution, patents and entrepreneurial self-efficacy.

In what follows we explain in detail a conceptual model, together with both dependent variables and individual factors that influence an academic's entrepreneurial involvement or an academic's intention to become an entrepreneur.

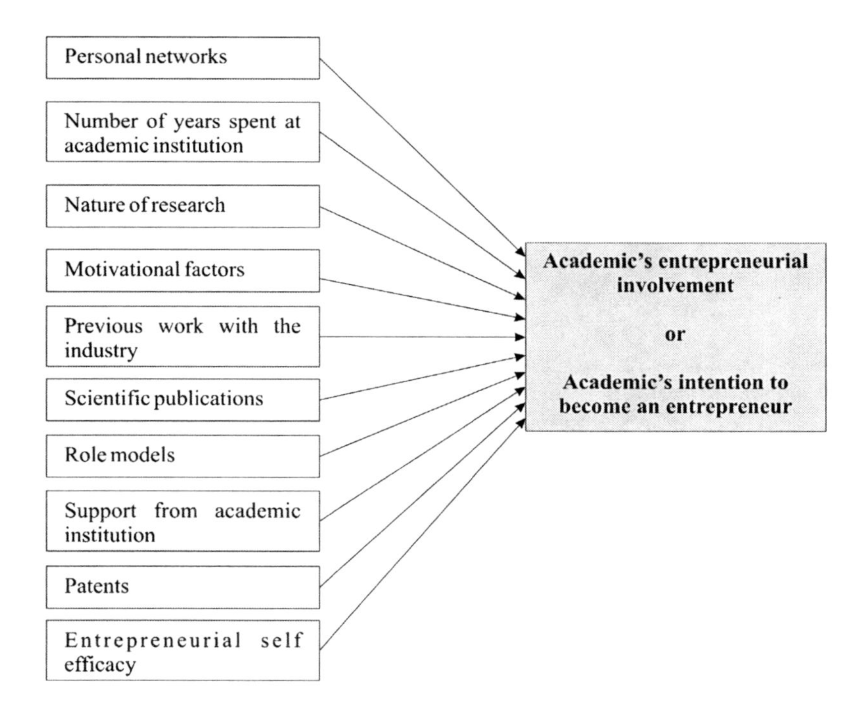

Figure 1: A Conceptual framework for studying technology transfer from academia to new firms from the academic/academic entrepreneur's point of view.

Dependent Variables

Academic's Entrepreneurial Involvement The academic entrepreneur needs to make choices in terms of committing full time to a spin-off or academic institution or working part-time at both. On the one hand, the academic may leave academia to completely focus his or her energy on a new firm; on the other hand, the inventor may decide to remain in the academia and may or may not accept a part time position in the new company (Nicolaou & Birley, 2003b). Harmon et al. (1997) found that few university inventors leave the university, but rather generally help to commercialize the invention on a part-time basis. When academics establish a spin-off company this does not necessarily mean that they leave their academic position permanently nor take a leave of absence (Goldfarb & Henrekson, 2003). Richter (1986) estimated that 3.3% of scientists and engineers who are employed full time as professors in American "four-year" higher educational institutions also work as consultants for commercial companies of which they are owners or part owners. Similarly, Allen and Norling (1991) found out that among 912 faculty members in science, engineering, business, and medicine, 16.2% of academics were engaged in firm formation, whereas only 4.4% were engaged in firm formation on the basis of their academic research.

Most scholars have tried to identify differences between entrepreneurs (academic entrepreneurs) and non-entrepreneurs (academics) for which they used dichotomous variable which coded entrepreneurs as 1 and non-entrepreneurs as 0, or they have tried to determine a typology of academic entrepreneurs. Since academic entrepreneurs are specific and since there are clearly also differences among those who have established a spin-off, we propose a new variable, called Academic's entrepreneurial involvement that measures the involvement of academic in his or her spin-off company. To our knowledge, so far there has been no scale variable that measured academic's entrepreneurial involvement. This type of dependent variable can then be used in regression models as dependent variable and also in structural equation modeling.

Intention to Become an Entrepreneur Intentions are the single best predictor of any planned behavior, including entrepreneurship (Krueger, Reilly, & Carsud, 2000). Intentions correspond to a state of mind that directs the individual's attention, experience, and action toward the goal of founding a business (Bird, 1988). Entrepreneurial intentions also suggest an individual's level of commitment to starting a new business (Krueger, 1993).

Independent Variables

Personal Networks Close relationships provide entrepreneurs with avenues for negotiation and persuasion, enabling them to gather various resources (e.g., market information, ideas, problem solving, social support, venture funding, and financial resources) held by other actors (Shane & Stuart, 2002; Hoang & Antoncic 2003; Nicolaou & Birley, 2003a; Nicolaou & Birley, 2003b; Walter et al., 2006). Nicolaou and Birley (2003a) in a literature review, defined business networks as those which can benefit from

- opportunity identification — the academic inventor is in an advantageous position to better identify market niches and may adapt his invention accordingly,
- access to important information and resources that could not otherwise be obtained,
- timing, where through business contacts the academic acquires early market information which can be of catalytic importance to research and development,
- receiving positive recommendations and evaluation at the right place through referrals. For example, venture capitalists and business angels are more inclined to invest in spin-offs that they know or that have been referred to them by reliable contacts, because this tends to alleviate informational asymmetry problems (Shane & Stuart, 2002).

Proposition 1. *There is a positive relationship between academic spin-off behavior (entrepreneurial involvement and intention to become an entrepreneur) and importance and frequency of interaction with persons with whom academic entrepreneurs/academics can discuss business matters.*

Number of Years Spent at the Academic Institution Most members of the academic community have through a tenured professorship guaranteed their socio–economic status, which provides a basis for spin-off creation. Their job stability and academic reputation normally are dependent upon teaching and publications. Without taking sufficient precautions, a faculty member may jeopardize his or her academic career by engaging in spin-off creation while shirking basic research responsibilities (Lee & Gaertner, 1994). Thus, the number of years spent at an academic institution is a proxy for their scientific seniority, which should negatively affect the level of academic's entrepreneurial involvement and intention to become an entrepreneur.

Proposition 2. *There is a negative relationship between the number of years spent at the academic institution and academic spin-off behavior (entrepreneurial involvement and intention to become an entrepreneur).*

The Nature of Research In general, academic research is oriented more towards basic research, which is driven by a scientist's curiosity or interest in a scientific question, rather than applied research. Basic research is experimental or theoretical work undertaken primarily to acquire new knowledge of the underlying foundation of phenomena and observable facts, without any particular application or use in view (OECD, 2002). Applied research is also original investigation undertaken to acquire new knowledge. It is, however, directed primarily towards a specific practical aim or objective (OECD, 2002) with market potential and thus more interesting for commercialization than basic research.

Proposition 3. *There is a positive relationship between applied research and academic spin-off behavior (entrepreneurial involvement and intention to become an entrepreneur).*

Motivational Factors Impacting Academic Spin-off Behavior In a recent exploratory study at MIT, Shane (2004) uncovered motivational characteristics of academic entrepreneurs, such as a desire to bring technology to the market (Samson & Gurdon, 1993; Weatherston, 1993; Corman, Perles, & Vancini, 1988; Shane, 2004); a desire for wealth (Roberts, 1991; Shane, 2004), and a desire for independence (Roberts, 1991; Shane, 2004), as key pull and push factors impacting academic spin-off behavior. Besides those motivational characteristics of academic entrepreneurs that were discussed by Shane (2004), there are some other motivational factors that apply to technical entrepreneurs and were discussed by other scholars (e.g., Roberts, 1991), including doing something others could not, and taking on and meeting broader responsibilities. On the basis of our literature review, and our knowledge of academic entrepreneurs, we additionally propose three additional motivational factors that were not tested in the literature and that impact upon academic spin-off behavior. These are the desire to secure additional research funding, dissatisfaction with the academic environment and a desire to pursue technological perfection.

Proposition 4. *There is a positive relationship between different motivational factors (discussed earlier) and academic spin-off behavior (entrepreneurial involvement and intention to become an entrepreneur).*

Previous Work with Industry At the institutional level, previous research on university–industry relationships indicates that institutions with closer ties to industry generate a greater number of spin-offs and exhibit more entrepreneurial activity, such as faculty consulting with industry, faculty involvement in new firms, and faculty and university equity participation in start-up firms (Roberts & Malone, 1996; Cohen, Florida, Randazzese, & Walsh, 1998 cited in Powers & McDougall, 2005).

The same also applies at the individual level. Blumenthal, Campbell, Causino, and Seashore (1996) surveyed 2052 academics at 50 universities in the life science field and found that industry funded academics are more commercially productive than those who are not industry funded. Similarly Mansfield's (1995) study of 66 firms as well as 200 academic researchers found that, in the early stages of research projects, academics receive more government versus industrial founding, while as a project matured, industry funding began to grow and academics became more involved as industry consultants. Corman et al. (1988) found that 90% of entrepreneurs interviewed (i.e., 20 of 22) were deeply involved in technical consulting activity before and often after launching their own firms. The Kassicieh, Radosevich, and Umbarger (1996) study found significant differences between entrepreneurs and non-entrepreneurs in terms of situational variables such as the level of involvement in business activities outside the laboratory. Colyvas et al. (2002) found in their examination of 11 case studies from Columbia University and Stanford University, that in all but one case, the researchers involved in spin-off were members of a network of scientists that included industry professionals. In a single case in which there was no academic and industry scientist linkage, the technology was never transferred.

Proposition 5. *There is a positive relationship between academic contacts with the industry, and academic spin-off behavior (entrepreneurial involvement and intention to become an entrepreneur).*

Scientific Publications Academic institutions typically do not reward activities such as commercializing research and creating new spin-off firms in their promotional and tenure decisions (Siegel et al., 2003a). Thus, academics are usually more interested in publishing their results, presenting them at conferences, and continuing in the academic research race (Graft et al., 2002), rather than being involved in the patenting and commercialization of research. The academic reward structure encourages the production of knowledge that is a useful input to other academics' research. Researchers wish to have their papers cited because this is a signal that they have established a reputation within the academic community (Goldfarb & Henrekson, 2003), which is the primary motivation for university scientists (Siegel et al., 2003b). Different scholars have argued that publishing papers and striving for citations is a central objective of academic research, as citation measures are associated with higher income and prestige (e.g., Diamond, 1986; Stern, 2004) and also as a recognition from other scientists which may lead to election to a national academy and the ultimate accolade of a Nobel prize (Etzkowitz, 1998).

The performance evaluation process and publishing-orientated research thus act as barriers to creation of new academic spin-offs (Ndonzuau et al., 2002). There is little reason to believe that the goal of producing useful inputs to the research of other academics (which is done through scientific publications) is congruent with the goal of producing commercially valuable knowledge. Hence, efforts directed at the production of commercially valuable knowledge will most likely come at the expense of the production of a recognized reputation for the academic (Goldfarb & Henrekson, 2003).

The importance of a reward system in academic institutions as barrier to creation of new academic spin-off firms is also illustrated by the study by Siegel, Waldman, Atwater, and Link (2004) which was based on 55 structured interviews of three types of university-industry technology transfer stakeholders (managers/entrepreneurs, technology transfer office directors/university administrators, and university scientists). They found that 80% of managers/entrepreneurs and 85% of technology transfer office directors/university administrators and university scientists identified the importance of modifying the reward system in universities to reward technology transfer activities, as key to improving rates of university-industry technology transfer.

Proposition 6. *There is a negative relationship between publishing recognizable scientific papers and academic spin-off behavior (entrepreneurial involvement and intention to become an entrepreneur).*

Role Models The impact of role models on entrepreneurial behavior has been studied by many researchers and it has been found to correlate significantly with entrepreneurial behavior and intentions (Roberts, 1991; Krueger et al., 2000). Once a university has established an entrepreneurial tradition, and a number of successful companies, fellow faculty members can offer material support, in addition to moral support, to colleagues who are trying to establish a company of their own (Etzkowitz, 1998). Academics who have started their own firms can also become advisors to those newly embarking on a venture. The effort by pioneering faculty members in founding companies can lead other faculty members to found companies as well, because it leads the followers to believe that firm formation was an easy and desirable activity (Feldman, Feller, Bercovizt, & Burton, 2000 cited in Shane, 2004). Similarly in a large sample study (although based on case studies), Audretsch, Weigand, and Weigand (2000) provides similar results, showing that science-based firm formation is in fact, influenced by a demonstration effect of prior start-up efforts by other scientists. Similarly conclusions were made also by Shane (2004) and Etzkowitz (1998). Etzkowitz (1998) cited an aspiring academic entrepreneur that recalled that a department colleague who had formed a company, "gave me a lot of advice…he was the role model."

The availability of such role models makes it more likely that other academics will form a firm out of their research results, when the opportunity appears.

Proposition 7. *There is a positive relationship between availability of role models (academic entrepreneurs) and academic spin-off behavior (entrepreneurial involvement and intention to become an entrepreneur).*

Support from Academic Institution Lockett and Wright (2005) have argued that there is a positive relationship between incentives and rewards for establishing a university spin-offs and the creation of university spin-offs. Siegel et al. (2003b) found out that barriers to university-industry technology transfer are also found in university aggressiveness towards exercising intellectual property rights and bureaucracy and inflexibility of university administrators. Additionally, Degroof and Roberts (2004) have proposed that constructive spin-off policies in academic institutions significantly affect the growth potential of spin-off companies. Thus, if academics perceive support from academic institution, he or she will more likely become an entrepreneur or will easier be more involved as entrepreneur.

Proposition 8. *There is a positive relationship between support from academic institution and academic spin-off behavior (entrepreneurial involvement and intention to become an entrepreneur).*

Patents Patenting is a logical extension of the tendency toward increasing interest in commercially applicable results (Louis, Blumenthal, Gluck, & Stoto, 1989).

Proposition 9. *There is a positive relationship between number of patents (applied/ granted) of academic and academic spin-off behavior (entrepreneurial involvement and intention to become an entrepreneur).*

Entrepreneurial Self-Efficacy It is widely acknowledged that most scientist lack the business background needed to bring technology closer to the market (Druilhe & Garnsey, 2004) and many established spin-off companies can be characterized by a lack of commercial awareness that may lead the company to become technology rather than market driven. Typically, technology orientated entrepreneurs seeks to develop the absolute best "mousetrap" and constantly pursues perfection (Wilem, 1991). Products never sell themselves, and there is always the need for varying degrees of marketing and sales skills (Sljivic, 1993). The ability to connect technical knowledge and a commercial opportunity requires a set of skills, aptitudes, insights, and circumstances that are neither uniformly nor widely distributed (Venkataraman, 1997). Besides commercial knowledge, new academic entrepreneurs also require administrative skills, since where previously all the administration was done by the university, spin-off companies have to address these time consuming and distracting aspects themselves (Sljivic, 1993).

Thus, the creation of a new venture by academics can be described as a process in which they are involved in both the invention and the commercial exploitation phase (Grandi & Grimaldi, 2005). Thus, they need both specific scientific knowledge and also business related skills or at least certainty in performing business related roles and tasks. The certainty in performing business-related roles and tasks of entrepreneurs is entrepreneurial self-efficacy, which is relatively more general than task self-efficacy (Chen, Greene, & Crick, 1998). Entrepreneurial self-efficacy refers to the strength of an individual's belief that he or she is capable of successfully performing the roles and tasks of an entrepreneur (Boyd & Vozikis, 1994).

Proposition 10. *Academic entrepreneurs with perceived high entrepreneurial self-efficacy are more likely to be more involved in the spin-offs that they have established.*

Proposition 11. *Academics with perceived high entrepreneurial self-efficacy are more likely to become entrepreneurs than those with low entrepreneurial self-efficacy.*

Control Variables

Planned or Spontaneously Occurring Spin-Off Steffensen et al. (1999) identify two types of spin-offs; these are planned, when the new venture results from an organized effort by the parent organization, and spontaneously occurring, when the new company is established by an entrepreneur who identifies a market opportunity and who founds the spin-off with little encouragement (and perhaps with discouragement) from the parent organization. Since in planned spin-offs, academics are much more influenced by parent organization (academic institution) we will control for this variable.

Other Controlled Variables Such other controlled variables include gender, age, years since establishment of own company, total years of employment, percentage of equity in spin-off company of academic institution, percentage of academic's equity in spin-off company, whether an establish company arise from academic research. Number of entrepreneurs in establishing a spin-off and highest professional degree attained at the academic institution (researcher, doctoral researcher, post-doctoral research associate, assistant professor, associate professor, full professor, other).

Conclusions

To conclude this literature review of a conceptual framework for studying a technology transfer process and our understanding of technology transfer from academia to new firms, we propose a main research thesis as follows. Academic spin-off behavior (entrepreneurial involvement and intention to become an entrepreneur) is from entrepreneur's standpoint influenced by entrepreneurial self-efficacy, personal motivational factors, the nature of the research (e.g., basic versus applied research), number of years spent at the academic institution, patenting, availability of personal networks of academics or academics entrepreneurs, previous work with the industry, availability of role models, publishing recognizable scientific papers, and support from academic institution.

We believe that the proposed conceptual framework for studying technology transfer will help researchers, policy makers, and practitioners in designing policy measures and instruments to foster technology transfer from academia to new firms.

References

Acs, Z. J., Audretsch, D. B., & Feldman, M. P. (1992). Real effects of academic research: Comment. *American Economic Review*, *82*(1), 363–367.

Allen, D. N., & Norling, F. (1991). Exploring perceived threats in faculty commercialization of research. In: B. M. Alistair, R. W. Smilor & D. V. Gibson (Eds), *University spin-off companies: Economic development, faculty entrepreneurs, and technology transfer*. Lanham: Rowman and Littlefield Publishers.

Association of University Technology Managers. (2002). *The AUTM licensing survey: Fiscal year 2000. Survey summary*. Norwalk: Association of University Technology Managers.

Audretsch, D. B., Weigand, J., & Weigand, C. (2000). Does the small business innovation research program foster entrepreneurial behavior? Evidence from Indiana. In: C. W. Wessner (Ed.), *The small business innovation research program: An assessment of the department of defense fast track initiative*. Washington: National Academy Press.

BankBoston. (1997). *MIT: the impact of innovation*. Boston: BankBoston.

Bird, B. (1988). Implementing entrepreneurial ideas: The case for intention. *Academy of Management Review*, *13*(3), 442–453.

Blumenthal, D., Campbell, E. G., Causino, N., & Seashore, L. K. (1996). Participation of life-science faculty in research relationships with industry. *New England Journal of medicine*, *335*(23), 1734–1739.

Boyd, N. G., & Vozikis, G. S. (1994). The influence of self-efficacy on the development of entrepreneurial intentions and actions. *Entrepreneurship: Theory and Practice*, *18*(4), 63–77.

Brennan, M. C., & McGowan, P. (2007). The knowledge market place: Understanding interaction at the academic-industry interface. In: J. Ulijn, D. Drillon & F. Lasch (Eds), *Entrepreneurship, co-operation and the firm: The emergence and survival of high tech ventures in Europe*. Cheltenham, UK and Lyme, US: Edward Elgar.

Carayannis, E. G., Rogers, E. M., Kurihara, K., & Allbritton, M. M. (1998). High-technology spin-offs from government R&D laboratories and research universities. *Technovation*, *18*(1), 1–11.

Chapple, W., Lockett, A., Siegel, D., & Wright, M. (2005). Assessing the relative performance of U.K. university technology transfer offices: Parametric and non-parametric evidence. *Research Policy*, *34*(3), 369–384.

Chen, C. C., Greene, P. G., & Crick, A. (1998). Does entrepreneurial self-efficacy distinguish entrepreneurs from managers? *Journal of Business Venturing*, *13*(4), 295–316.

Chiesa, V., & Piccaluga, A. (1998). Transforming rather than transferring scientific and technological knowledge-the contribution of academic spin-out companies: The Italian way. In: R. Oakey & W. During (Eds), *New technology-based firms in the 1990s* (vol. 4). London: Paul Chapman.

Clarysse, B., Wright, M., Lockett, A., Van de Velde, E., & Vohora, A. (2005). Spinning out new ventures: A typology of incubation strategies from European research institutions. *Journal of Business Venturing*, *20*(2), 183–216.

Cohen, W. M., Florida, R., Randazzese, L., & Walsh, J. (1998). Industry and the academy: Uneasy partners in the cause of technological advance. In: R. Noll (Ed.), *Challenge to the research university*. Washington: Brookings Institution.

Colyvas, J., Crow, M., Gelijns, A., Mazzoleni, R., Nelson, R. R., Rosenberg, N., & Sampat Bhaven, N. (2002). How do university inventions get into practice? *Management Science*, *48*(1), 61–72.

Cooper, A. C. (1971). Spin-offs and technical entrepreneurship. *I.E.E.E. Transactions on Engineering Management*, *18*(1), 2–6.

Corman, J., Perles, B., & Vancini, P. (1988). Motivational factors influencing high-technology entrepreneurship. *Journal of Small Business Management*, *26*(1), 36–42.

Debackere, K., & Veugelers, R. (2005). The role of academic technology transfer organizations in improving industry science links. *Research Policy*, *34*(3), 321–342.

Degroof, J. J., & Roberts, E. B. (2004). Overcoming weak entrepreneurial infrastructures for academic spin-off ventures. *The Journal of Technology Transfer*, *29*(3-4), 327–352.

Diamond, A. M. (1986). What is a citation worth? *Journal of Human Resources*, *21*(2), 200–215.

Druilhe, C., & Garnsey, E. (2004). Do academic spin-outs differ and does it matter? *The Journal of Technology Transfer*, *29*(3-4), 269–285.

Etzkowitz, H. (1998). The norms of entrepreneurial science: Cognitive effects of the new university-industry linkages. *Research Policy*, *27*(8), 823–833.

Etzkowitz, H. (2003). Research groups as 'quasi-firms': The invention of the entrepreneurial university. *Research Policy*, *32*(1), 109–121.

Feldman, M., Feller, I., Bercovizt, J., & Burton, R. (2000). *Understanding evolving university-industry relationships*. Working paper, Johns Hopkins University.

Franklin, S. J., Wright, M., & Lockett, A. (2001). Academic and surrogate entrepreneurs in university spin-out companies. *Journal of Technology Transfer*, *26*(1-2), 127–141.

Garvin, D. A. (1983). Spin-offs and the new firm formation process. *California Management Review*, *25*(2), 3–20.

Goldfarb, B., & Henrekson, M. (2003). Bottom-up versus top-down policies towards the commercialization of university intellectual property. *Research Policy*, *32*(4), 639–658.

Goldfarb, B., Henrekson, M., & Rosenberg, N. (2001). *Demand vs. supply driven innovations: US and Swedish experiences in academic entrepreneurship*. Working Paper Series in Economics and Finance 0436, Stockholm School of Economics.

Graft, G., Heiman, A., & Zilberman, D. (2002). University research and offices of technology transfer. *California Management Review*, *45*(1), 88–115.

Grandi, A., & Grimaldi, R. (2005). Academics' organizational characteristics and the generation of successful business ideas. *Journal of Business Venturing*, *20*(6), 821–845.

Gregorio, D. D., & Shane, S. (2003). Why do some universities generate more start-ups than others? *Research Policy*, *32*(2), 209–227.

Harmon, B., Ardishvili, A., Cardozo, R., Elder, T., Leuthold, J., Parshall, J., Raghian, M., & Smith, D. (1997). Mapping the university technology transfer process. *Journal of Business Venturing*, *12*(6), 423–434.

Hoang, H., & Antoncic, B. (2003). Network-based research in entrepreneurship: A critical review. *Journal of Business Venturing*, *18*(2), 165–187.

Jaffe, A. B. (1989). Real effects of academic research. *American Economic Review*, *79*(5), 957–970.

Jensen, R., & Thursby, M. (2001). Proofs and prototypes for sale: The licensing of university inventions. *The American Economic Review*, *91*(1), 240–259.

Jensen, R. A., Thursby, J. G., & Thursby, M. C. (2003). Disclosure and licensing of university inventions: 'the best we can do with the s**t we get to work with. *International Journal of Industrial Organization*, *21*(9), 1271–1300.

Kassicieh, S. K., Radosevich, R. H., & Umbarger, J. (1996). A comparative study of entrepreneurship incidence among inventors in national laboratories. *Entrepreneurship: Theory and practice*, *20*(3), 33–49.

Krueger, N. (1993). The impact of prior entrepreneurial exposure on perceptions of new venture feasibility and desirability. *Entrepreneurship: Theory and Practice*, *18*(1), 5–21.

Krueger, N. F., Reilly, M. D., & Carsud, A. L. (2000). Competing models of entrepreneurial intentions. *Journal of Business Venturing*, *15*(5-6), 411–432.

Lee, Y., & Gaertner, R. (1994). Technology transfer from university to industry: A large-scale experiment with technology development and commercialization. *Policy Studies Journal*, *22*(2), 384–401.

Leitch, C. M., & Harrison, R. T. (2005). Maximizing the potential of university spin-outs: The development of second-order commercialization activities. *R&D Management*, *35*(3), 257–272.

Levin, S. G., & Stephan, P. E. (1991). Research productivity over the life cycle: Evidence for academic scientists. *American Economic Review*, *81*(1), 114–132.

Liu, H., & Jiang, Y. (2001). Technology transfer from higher education institutions to industry in China: Nature and implications. *Technovation*, *21*(3), 175–188.

Lockett, A., Siegel, D., Wright, M., & Ensley, M. D. (2005). The creation of spin-off firms at public research institutions: Managerial and policy implications. *Research Policy*, *34*(7), 981–993.

Lockett, A., & Wright, M. (2005). Resources, capabilities, risk capital and the creation of university spin-out companies. *Research Policy*, *34*(7), 1043–1057.

Louis, K. S., Blumenthal, D., Gluck, M. E., & Stoto, M. A. (1989). Entrepreneurs in Academe: An exploration of behaviors among life scientists. *Administrative Science Quarterly*, *34*(1), 110–131.

Mansfield, E. (1991). Academic research and industrial innovation. *Research Policy*, *20*(1), 1–12.

Mansfield, E. (1995). Academic research underlying industrial innovations: Sources, characteristics, and financing. *Review of Economics and Statistics*, *77*(1), 55–65.

Mansfield, E. (1998). Academic research and industrial innovation: An update of empirical findings. *Research Policy*, *26*(7-8), 773–776.

Markman, G. D., Gianiodis, P. T., Phan, P. H., & Balkin, D. B. (2005b). Innovation speed: Transferring university technology to market. *Research Policy*, *34*(7), 1058–1075.

Melkers, J., Bulger, D., & Bozeman, L. (1993). Technology transfer and economic development. In: R. Bingham & R. Mier (Eds), *Theories of local economic development*. Newbury Park: Sage.

Mian, S. A. (1997). Assessing and managing the university technology business incubator: An integrative framework. *Journal of Business Venturing*, *12*(4), 251–285.

Mustar, P. (1997). Spin-off enterprises-how French academics create hi-tech companies: The conditions for success or failure. *Science and Public Policy*, *24*(1), 37–43.

Ndonzuau, F. N., Pirnay, F., & Surlemont, B. (2002). A stage model of academic spin-off creation. *Technovation*, *22*(5), 281–289.

Nicolaou, N., & Birley, S. (2003a). Academic networks in a trichotomous categorization of university spinouts. *Journal of Business Venturing*, *18*(3), 333–359.

Nicolaou, N., & Birley, S. (2003b). Social networks in organizational emergence: The university spinout phenomenon. *Management Science*, *49*(12), 1702–1725.

OECD. (2002). *Frascati manual: Proposed standard practice for surveys on research and experimental development*. Paris: OECD.

O'Shea, R. P., Allen, T. J., Chevalier, A., & Roche, F. (2005). Entrepreneurial orientation, technology transfer and spinoff performance of U.S. universities. *Research Policy*, *34*(7), 994–1009.

Perez, M. P., & Sanchez, A. M. (2003). The development of university spin-offs: Early dynamics of technology transfer and networking. *Technovation*, *23*(10), 823–831.

Phillips, R. G. (2002). Technology business incubators: How effective as technology transfer mechanisms? *Technology in Society*, *24*(3), 299–316.

Pirnay, F., Surlemont, B., & Frederic, N. (2003). Toward a typology of university spin-offs. *Small Business Economics*, *21*(4), 355–369.

Powers, J. B., & McDougall, P. P. (2005). University start-up formation and technology licensing with firms that go public: A resource-based view of academic entrepreneurship. *Journal of Business Venturing*, *20*(3), 291–311.

Radosevich, R. (1995). A model for entrepreneurial spin-offs from public technology sources. *International Journal of Technology Management*, *10*(7-8), 879–893.

Rahm, D. (1994). Academic perceptions of university-firm technology-transfer. *Policy Studies Journal*, *22*(2), 267–278.

Rappert, B., Webster, A., & Charles, D. (1999). Making sense of diversity and reluctance: Academic-industrial relations and intellectual property. *Research Policy*, *28*(8), 873–890.

Richter, M. N. (1986). University scientists as entrepreneurs. *Society*, *23*(5), 81–83.
Roberts, E. B. (1991). *Entrepreneurs in high technology: Lessons from MIT and beyond*. Oxford: Oxford University Press.
Roberts, E. B., & Malone, D. E. (1996). Policies and structures for spinning off new companies from research and development organizations. *R&D Management*, *26*(1), 17–48.
Rogers, E. M., & Takegami Shiro, Y. J. (2001). Lessons learned about technology transfer. *Technovation*, *21*(4), 253–261.
Samson, K., & Gurdon, M. (1993). University scientists as entrepreneurs: A special case of technology transfer and high-tech venturing. *Technovation*, *13*(2), 63–71.
Shane, S. (2001a). Technological opportunities and new firm creation. *Management Science*, *47*(2), 205–220.
Shane, S. (2001b). Technology regimes and new firm formation. *Management Science*, *47*(9), 1173–1190.
Shane, S. (2002). Selling university technology: Patterns from MIT. *Management Science*, *48*(1), 122–138.
Shane, S. (2004). *Academic entrepreneurship: University spinoffs and wealth creation*. Cheltenham: Edward Elgar Publishing.
Shane, S., & Stuart, T. (2002). Organizational endowments and the performance of university start-ups. *Management Science*, *48*(1), 154–170.
Siegel, D. S., Waldman, D., & Link, A. (2003a). Assessing the impact of organizational practices on the relative productivity of university technology transfer offices: An exploratory study. *Research Policy*, *32*(1), 27–48.
Siegel, D. S., Waldman, D. A., Atwater, L. E., & Link, A. N. (2003b). Commercial knowledge transfers from universities to firms: Improving the effectiveness of university-industry collaboration. *The Journal of High Technology Management Research*, *14*(1), 111–133.
Siegel, D. S., Waldman, D. A., Atwater, L. E., & Link, A. N. (2004). Toward a model of the effective transfer of scientific knowledge from academicians to practitioners: Qualitative evidence from the commercialization of university technologies. *Journal of Engineering and Technology Management*, *21*(1-2), 115–142.
Sljivic, N. (1993). University spin-off companies: Management requirements and pitfalls to be avoided. *International Journal of Educational Management*, *7*(5), 32–34.
Smilor, R. W., Gibson, D. V., & Dietrich, G. B. (1990). University spin-out companies: Technology start-ups from UT-Austin. *Journal of Business Venturing*, *5*(1), 63–76.
Steffensen, M., Rogers, E. M., & Speakman, K. (1999). Spin-offs from research centers at a research university. *Journal of Business Venturing*, *15*(1), 93–111.
Stern, S. (2004). Do scientists pay to be scientists? *Management Science*, *50*(6), 835–853.
Storper, M. (1995). The resurgence of regional economics, ten years later: The region as a nexus of untraded interdependencies. *European Urban and Regional Studies*, *2*(3), 191–221.
Thursby, J. G., Jensen, R., & Thursby, M. C. (2001). Objectives, characteristics and outcomes of university licensing: A survey of major U.S. universities. *Journal of Technology Transfer*, *26*(1-2), 59–72.
Venkataraman, S. (1997). The distinctive domain of entrepreneurship research. In: J. A. Katz (Ed.), *Advances in entrepreneurship, firm emergence and growth (Vol. 3)*. Greenwich: AI Press.
Walsh, S. T., & Kirchhoff, B. A. (2002). Technology transfer from government labs to entrepreneurs. *Journal of Enterprising Culture*, *10*(2), 133–149.
Walter, A., Auer, M., & Ritter, T. (2006). The impact of network capabilities and entrepreneurial orientation on university spin-off performance. *Journal of Business Venturing*, *21*(4), 541–567.

Weatherston, J. (1993). Academic entrepreneurs. *Industry and Higher Education*, *7*(December), 235–243.

Weatherston, J. (1995). *Academic entrepreneurs: Is a spin-off company too risky*? Proceedings of the 40th International Council on Small Business, Sydney, 18–21 June.

Wilem, F. J. (1991). The breeder: Forming spin-off corporations through university-industry partnerships. In: B. M. Alistair, R. W. Smilor & D. V. Gibson (Eds), *University spin-off companies: Economic development, faculty entrepreneurs, and technology transfer*. Lanham: Rowman & Littlefield Publishers.

Wright, M., Birley, S., & Mosey, S. (2004). Entrepreneurship and university technology transfer. *The Journal of Technology Transfer*, *29*(3-4), 235–246.

Zucker, L. G., Darby, M. R., & Brewer, M. B. (1998). Intellectual human capital and the birth of U.S. biotechnology enterprises. *American Economic Review*, *88*(1), 290–306.